# Collaborative Learning

## Cognitive and Computational Approaches

# ADVANCES IN LEARNING AND INSTRUCTION SERIES

Series Editors:
Neville Bennett, Erik DeCorte, Stella Vosniadou and Heinz Mandl

**Published**

VAN SOMEREN, REIMANN, BOSHUIZEN & DE JONG
Learning with Multiple Representations

**Forthcoming titles**

COWIE, AALSVOORT & MERCER
Social Interaction in Learning and Instruction

KAYSER & VOSNIADOU
Modelling Changes in Understanding: Case Studies in Physical Reasoning

ROUET
Using Complex Information Systems

SÄLJÖ, BLISS & LIGHT
Learning Sites: Social and Technological Resources for Learning

SCHNOTZ, VOSNIADOU & CARRETERO
New Perspectives on Conceptual Change

**Other titles of interest**

REIMANN & SPADA
Learning in Humans and Machines: Towards an Interdisciplinary Learning Science

Computer Assisted Learning: Proceedings of the CAL series of biennial Symposia 1989, 1991, 1993, 1995 and 1997 (five volumes)

**Related journals - sample copies available on request**

Learning and Instruction
International Journal of Educational Research
Computers and Education
Computers and Human Behavior

# Collaborative Learning

## Cognitive and Computational Approaches

*edited by*

*Pierre Dillenbourg*



ELSEVIER

Amsterdam • Boston • Heidelberg • London
New York • Oxford • Paris • San Diego
San Francisco • Singapore • Sydney • Tokyo
Earli is an imprint of Elsevier

Earli

ELSEVIER SCIENCE Ltd
The Boulevard, Langford Lane
Kidlington, Oxford OX5 1GB, UK



First edition 1999
Second impression 2003

Library of Congress Cataloging-in-Publication Data

Collaborative learning : cognitive and computational approaches /
edited by Pierre Dillenbourg.
p. cm. -- (Advances in learning and instruction series)
Includes bibliographical references and indexes.
ISBN 0-08-043073-2 (hc.)
1. Group work in education. 2. Team learning approach in education. 3. Cognitive science. 4. Learning, Psychology of.
I. Dillenbourg, Pierre. II. Series.
LB1032.C566 1999
371.39'5--dc21 98-38305
CIP

British Library Cataloguing in Publication Data
A catalogue record from the British Library has been applied for.

ISBN: 0 08 043073 2

♾ The paper used in this publication meets the requirements of ANSI/NISO Z39.48-1992 (Permanence of Paper).
Printed in Hungary.

# Table of Contents


*Acknowledgement* vii
*Contributors* ix

1. Introduction; What Do You Mean By "Collaborative Learning"?
*Pierre Dillenbourg* 1

2. Learning Together: Understanding the Processes of Computer-Based Collaborative Learning
*Karen Littleton and Päivi Häkkinen* 20

3. The Role of Grounding in Collaborative Learning Tasks
*Michael Baker, Tia Hansen, Richard Joiner and David Traum* 31

4. What is "Multi" in Multi-agent Learning?
*Gerhard Weiss and Pierre Dillenbourg* 64

5. Comparing Human–Human and Robot–Robot Interactions
*Richard Joiner, Kim Issroff and John Demiris* 81

6. Learning by Explaining to Oneself and to Others
*Rolf Ploetzner, Pierre Dillenbourg, Michael Preier and David Traum* 103

7. Knowledge Transformations in Agents and Interactions: A Comparison of Machine Learning and Dialogue Operators
*Engelbert Mephu Nguifo, Michael Baker and Pierre Dillenbourg* 122

8. Can Analytic Models Support Learning in Groups?
*H. Ulrich Hoppe and Rolf Ploetzner* 147

9. Using Telematics for Collaborative Knowledge Construction
*Tia Hansen, Lone Dirckinck-Holmfeld, Robert Lewis and Jože Rugelj* 169

10. The Productive Agency that Drives Collaborative Learning
*Daniel L. Schwartz* 197

*References* 219

*Index* 243

# Acknowledgement

Producing this book would not have been possible without the support of the European Science Foundation, Strasbourg. The European Science Foundation is an association of 62 major national funding agencies devoted to basic scientific research in 21 countries. The ESF assists its Member Organisations in two main ways: by bringing scientists together in its Scientific Programmes, Networks and European Research Conferences, to work on topics of common interest; and through the joint study of issues of strategic importance in European science policy. The scientific work sponsored by the ESF includes basic research in the natural and technical sciences, the medical and biosciences, the humanities and social sciences. The ESF maintains close relations with other scientific institutions within and outside Europe. Through its activities, the ESF adds value by cooperation and coordination across national frontiers and endeavours, offers expert scientific advice on strategic issues, and provides the European forum for fundamental science. This book is one of the outcomes of the ESF Scientific Programme on "Learning in Humans and Machines".

# Contributors

Michael Baker
UMR GRIC–COAST
CNRS & Université Lumière Lyon 2
5, Avenue Pierre-Mendes-France
69676 Bron Cedex
France
*email*: mbaker@univ-lyon2.fr

John Demiris
Division of Informatics
University of Edinburgh
5 Forrest Hill
EH1 2QL
UK
*email*: johnde@dai.ed.ac.uk

Pierre Dillenbourg
TECFA
Faculty of Education and Psychology
Universite de Geneve
9, route de Drize
1227 Carouge
Switzerland
*email*: Pierre.Dillenbourg@tecfa.unige.ch

Lone Dirckinck-Holmfeld
Institute of Communication
Aalborg University
Langagervej 8
DK-92220 Aalborg
Denmark
*email*: lone@hum.auc.dk

Päivi Häkkinen
Institute for Educational Research
University of Jvyäskylä
PO Box 35
40351 Jyväskylä
Finland
*email*: phakkine@cc.jyu.fi

Tia Hansen
Institute of Psychology
University of Århus
Asylvej 4
DK-8240 Risskov
Denmark
*email*: tia@psy.au.dk

H. Ulrich Hoppe
Gerhard-Mercator-University GH Duisburg
Faculty Mathematics/Department Informatics
Lotharstr. 63/65; Building LF
D-47048 Duisburg
Germany
*email*: hoppe@informatik.uni-duisburg.de

Kim Issroff
Higher Education Research and Development Unit
University College London
1–19 Torrington Place
London
WC1E 6BT
UK
*email*: k.issroff@ucl.ac.uk

Richard Joiner
Faculty of Education
The Open University
Walton Hall
Milton Keynes
Buckinghamshire
MK7 6AA
UK
*email*: R.W.Joiner@Open.ac.uk

Robert Lewis
Knowledge Technology Research Unit
Information Systems
Library Building
Lancaster
LA1 4YW
UK
*email*: r.lewis@lancaster.ac.uk

Karen Littleton
Department of Psychology
Faculty of Social Sciences
The Open University
Walton Hall
Milton Keynes
MK7 6AA
UK
*email*: k.s.littleton@open.ac.uk

Engelbert Mephu Nguifo
CRIL
IUT de Lens—Université d'Artois
rue de l'Universite SP 16
62307 LENS Cedex
France
*email*: mephu@cril.univ-artois.fr

Rolf Ploetzner
University of Freiburg
Department of Psychology
Niemensstr. 10
D-79085 Freiburg
Germany
*email*: ploetz@psychologie.uni-freiburg.de

Michael Preier
University of Freiburg
Department of Psychology
Niemensstr. 10
D-79085 Freiburg
Germany

Jože Rugelj
University of Ljubljana
Faculty of Education
Kardeljeva pl. 16
1000 Ljubljana
Slovenia
*email*: joze.rugelj@uni-lj.si

Daniel L. Schwartz
Learning Technology Center
Box-45GPC
Vanderbilt University
Nashville, TN 37203
USA
*email*: schwardl@ctrvax.vanderbilt.edu

David Traum
UMIACS
University of Maryland
A. V. Williams Building
College Park
Maryland
USA
*email*: traum@cs.umd.edu

Gerhard Weiss
Institut für Informatik
Technische Universität München
Germany
*email*: weissg@informatik.tu-muenchen.de

# 1

# Introduction: What Do You Mean By "Collaborative Learning"?

Pierre Dillenbourg

## 1 Introduction

This book arises from a series of workshops on collaborative learning that gathered together 20 scholars from the disciplines of psychology, education and computer science. The series was part of a research programme entitled "Learning in Humans and Machines" (LHM), launched by Peter Reimann and Hans Spada, and funded by the European Science Foundation (ESF). This programme aimed to develop a multidisciplinary dialogue on learning, involving mainly scholars from cognitive psychology, educational science and artificial intelligence (including machine learning). During the preparation of the programme, Agnes Blaye, Claire O'Malley, Michael Baker and I developed a theme on collaborative learning. When the programme officially began, 12 members were selected to work on this theme and formed the so-called "Task Force 5". I became the coordinator of the group. This group organised two workshops, in Sitges (Spain, 1994) and Aix-en-Provence (France, 1995). In 1996, the group was enriched with new members to reach its final size. Around 20 members met in the subsequent workshops, at Samoens (France, 1996), Houthalen (Belgium, 1996) and Mannheim (Germany, 1997). Several individuals joined the group for some time but have not written a chapter in this book. I would nevertheless like to acknowledge their contributions to our activities: George Bilchev, Stevan Harnad, Calle Jansson and Claire O'Malley.

The reader will not be surprised to learn that our group did not agree on any definition of collaborative learning. We did not even try. There is such a wide variety of uses of this term inside each academic field and, *a fortiori*, across the fields. Moreover, the task force activities did not include a specific task that would have forced members to converge on a shared understanding of "collaborative learning". Instead, we shared a broad interest in multidisciplinary interactions, which was reified in this book.

Hence, I review the variety of approaches to "collaborative learning" that were present in our group. When a word becomes fashionable—as is the case with "collaboration"—it is often used abusively for more or less anything. The problem with such an over-general usage is two-fold. First, it is nonsense to talk about the cognitive effects ("learning") of "collaborative" situations if any situation can be labelled "collaborative". Second, it is difficult to articulate the contributions of various authors who use the same word very differently. Therefore, I explore various aspects of this definition, not in order to establish "the" correct definition of collaborative learning, but in order to help the reader to put the different chapters in perspective. I will not review the chapters here, but the reader might prefer to read them before reading this introduction.

The broadest (but unsatisfactory) definition of "collaborative learning" is that it is a *situation* in which *two or more* people *learn* or attempt to learn something *together*. Each element of this definition can be interpreted in different ways:

- "two or more" may be interpreted as a pair, a small group (3–5 subjects), a class (20–30 subjects), a community (a few hundreds or thousands of people), a society (several thousands or millions of people) . . . and all intermediate levels;
- "learn something" may be interpreted as "follow a course", "study course material", "perform learning activities such as problem solving", "learn from lifelong work practice", etc.;
- "together" may be interpreted as different forms of interaction: face-to-face or computer-mediated, synchronous or not, frequent in time or not, whether it is a truly joint effort or whether the labour is divided in a systematic way.

These three elements of the definition define the space of what is encountered under the label "collaborative learning": pairs learning through intensive synchronous joint problem solving during 1 or 2 hours, groups of students using electronic mail during a 1-year course, communities of professionals developing a specific culture across generations, etc. I explore this space along three dimensions: the scale of the collaborative situation (group size and time span), what is referred to as "learning" and what is referred to as "collaboration".

## 2 The Variety of Scales

The different situations mentioned above create objects of study with different *scales*: from 2 to many subjects, from 20 minutes to many years. For instance, most empirical research on the effectiveness of collaborative learning was concerned with a small scale: two to five subjects collaborating for 1 hour or so. At the opposite end of this scale, the label "computer-supported collaborative learning" (CSCL) is often applied to situations in which a group of 40 subjects follows a course over 1 year. The findings of the former cannot, of course, be generalised to predict the outcomes of the latter and vice versa.

Generalisability is not only reduced by the difference in empirical settings, but also it has been affected by divergent underlying theories. Just as a photographer uses different lenses for photographing a flower or a mountain, scholars need different theoretical tools in order to grasp phenomena on various scales. While psychology provides useful frameworks for analysing learning in small groups, social psychology applies better to broader groups, and tools from sociology, ethnology or anthropology become relevant for larger scales. Even within psychology, different schools of thought have focused on different scales, as Karen Littleton and Päivi Häkkinen explain in chapter 2. These schools grew in geographically distinct places, namely Switzerland and Russia, at a time when distance and language were significant obstacles to scientific communication. They developed their own paradigm for empirical research that made the integration of results or concepts even more difficult. However, chapter 2 demonstrates the continuity between these different schools of thought and articulates them with current approaches.

The notion of scale has been the "Berlin Wall" of collaborative learning: for many years, it has compartmentalised the field, but it fell in the late 1980s:

> ". . . research paradigms built on supposedly clear distinctions between what is social and what is cognitive will have an inherent weakness, because the causality of social and cognitive processes is, at the very least, circular and is perhaps even more complex." (Perret-Clermont, Perret & Bell, 1991, p. 50)

This evolution is illustrated by the emergence of distributed cognition theories (Salomon, 1993) in which the group is viewed as a cognitive system. Scholars study objects with different scales, but theoretical concepts are imported across different scales. For instance, Hutchins (1995) describes group processes with the terms that cognitive science uses for describing individual cognition (memory, propagation of representational states, etc.). Conversely, the notion of culture that implicitly refers to the level of community or society (even in adult–child interactions, the adult's role was mainly to be a mediator of culture of his or her society, not the creator of this culture) is now applied to describe the common grounds built by peers in interaction:

> "The question is not how individuals become members in a larger cognitive community as they do in apprenticeship studies. Rather the question is how a cognitive community could emerge in the first place." (Schwartz, 1995, p. 350)

The process of building a group micro-culture is studied by Baker, Hansen, Joiner and Traum in chapter 3, and by Hansen, Dirckinck-Holmfeld, Lewis and Rugelj in chapter 9.

During their transfer across different scales, these concepts undergo deformations. For instance, the notion of group memory has not been elaborated as much as the notion of individual memory. In CSCL research, group memory is often reduced to a working memory, including a persistent representation of the problem state, mediated by some artefact (e.g. a shared visual workspace in groupware, an altitude meter in a cockpit, etc.) and an interaction memory (e.g. the trace of the last *n* interactions in a MUD environment). Conversely, if one talks about the culture built by two subjects after 1 hour of interaction, the term "culture" acquires a functional flavour (the subjects develop the language they need in order to solve a task) rather than its traditional historical flavour. This functional flavour can be perceived in chapter 3, which describes the grounding mechanisms, i.e. the mechanisms for co-constructing a common language. These mechanisms are subordinated to a functional criterion: two partners cannot gain perfect mutual understanding about everything in life; they simply need to have enough mutual understanding so that they are able to continue performing the task at hand (Clark & Brennan, 1991).

While distributed cognition treats the group as a single cognitive system, one may reciprocally view the individual as a distributed system (Minsky, 1987). This perspective broadens the scale issue, now including groups inside a single agent. Although it may sound awkward to talk about "collaboration with oneself", it common to talk about "conflict with oneself". The idea that thinking can be viewed as a dialogue with oneself is not a new idea; it has been argued by Piaget, Mead and, of course, Vygostky, for whom thought results from internalised dialogues. The relation between dialogue with oneself and dialogue with a peer is addressed by two chapters in this book:

- In chapter 6, Ploetzner, Dillenbourg, Preier and Traum compare explaining to oneself and explaining to somebody else. Learning by explaining to oneself received great attention in cognitive science, both in machine learning under the label "explanation-based learning" (Mitchell, Keller & Kedar-Cabelli, 1986) and in cognitive modelling under the label "the self-explanation effect" (VanLehn, Jones & Chi, 1992). In both models, explaining consists of building some proof in the artificial intelligence (AI) sense, i.e. to "understand" computationally. On the other hand, empirical research has established the cognitive effects of both (elaborated) self-explanations (Chi et al., 1989) and (elaborated) explanations (Webb, 1989, 1991). Does this imply that one can consider self-explanation and explanation as similar processes? Reviewing the literature on this issue, the authors of chapter 6 did not find any evidence that the interactivity of real explanation brings any benefit compared with self-explanation.
- In chapter 7, Mephu Nguifo, Baker and Dillenbourg address this issue at the computational level, by comparing the operators used to model the co-construction of knowledge through dialogue and those used in machine-learning research to model individual learning. Both sets of operators are rather similar at the knowledge level: for instance, "generalisation" can describe both the relation between two knowledge states during learning or the relationship between the semantic contents of two utterances in dialogue. However, this similarity does not extend to the strategy level: for instance, a dialogue strategy operator may be something like "lying to check one's partner's agreement", while a learning operator would be "focus on near-miss counter-examples".

These two chapters apparently overturn my expectations: in the comparison between dialogue with oneself and dialogue with a peer, the main difficulty might not be to identify similarities, but instead to establish exactly what differs between the two processes. Investigating when and how individual reasoning takes the form of a monologue could contribute to understanding the cognitive benefits of collaborative learning.

This evolution of research, where a group can be viewed as a unit or the individual as a group, indicates that the very notion of "scale" actually changes: it moves from a property of the object to a property of the observer, who selects the most appropriate unit of analysis. In computational models, the choice is rather open since there is no "natural" notion of agent as there is in psychology. A so-called "agent" can be any functional unit inside the system: an "edge detector" agent, an ANOVA agent, a grammatical parser agent, etc. Sometimes, a single rule is labelled as an agent; sometimes, it corresponds to an entire rulebase. One finds the same variety of "scales" in multi-agent systems as in psychology. Some systems include a few agents with elaborated skills while others include a large number of agents with elementary skills. The former perform meaningful computations, have goals, knowledge and even mutual representation. The latter are not viewed as intelligent agents, their interactions are not planned, but interesting phenomena hopefully emerge after a large number of interaction cycles. Sub-symbolic computation is often used in the latter. In chapter 4, Weiss and Dillenbourg describe some mechanisms of learning in multi-agent systems. Actually, in the term "multi-agent", we mainly discuss the prefix "multi", i.e. what is really specific to multiple agents in comparison with single-agent systems. I do not enter into the long debate in distributed artificial intelligence (DAI) regarding what can be called an agent. Agentship can be treated as a design metaphor. What must be assessed is whether the expectations that inevitably arise from such an anthropomorphic metaphor lead designers to be more productive or to cope with unsolved problems.

This book is mainly about the "small-scale" end of this continuum, i.e. collaboration between two or a few human or artificial agents for a well-defined learning or problem-solving task. This bias, initiated by the individual perspectives of most of the authors of the chapters in this book, has increased progressively through the series of workshops that we hold together. We deliberately left out of our debate some of the social and institutional factors that appear in large groups, such as leadership, the emergence of norms, and so forth. Chapters 3–7 focus on dyads and short learning periods. In chapter 8, Hoppe and Ploetzner describe a CSCL approach applicable to broader groups and curricula. However, the specificity of their approach is to identify sub-groups (peers) within the group, which would benefit from a collaborative interaction with respect to one curriculum item. The main criterion for matching learners is the complementarity of their skills or knowledge. Hope and Ploetzner present a computational model of the complementarity between quantitative and qualitative knowledge in physics. In other words, their focus is still on the narrow scale (two or three learners during 1 hour) within a large-scale system. This also characterises the first empirical study reported by Hansen, Dirckinck-Holmfeld, Lewis and Rugelj in chapter 9: within a broader institutional context, they study the interaction between two nurses and a physician. This second study is more illustrative of the "larger-scale" end of the spectrum. The authors consider a larger group, with a less-defined task, over a longer period of time. Not surprisingly, more institutional issues are raised during these observations. Specifically, learning is not studied in an instructional setting, but as personal and group development in work practices. This reveals a variety of understandings of the word "learning" in "collaborative learning" or the variety of tasks that are studied in collaborative learning research. I address this issue in the next section.

## 3 The Variety of Meanings For "Learning"

In the research literature on collaborative learning, there is a broad acceptance of what is put underneath the umbrella of "learning":

- For some scholars, it includes more or less any collaborative activity within an educational context, such as studying course material or sharing course assignments. The term "*collaborative learners*" would then be more appropriate.
- In other studies, and in most chapters of this book, the activity is joint problem solving, and learning is expected to occur as a side-effect of problem solving, measured by the elicitation of new knowledge or by the improvement of problem-solving performance. This understanding is also dominant in research on multi-agent learning (see chapter 4).
- Within some theories (see chapter 2), collaborative learning is addressed from a developmental perspective, as a biological and/or cultural process that occurs over years.
- This spectrum also includes learning from collaborative work, which refers to the lifelong acquisition of expertise within a professional community (see chapter 9).

In other words, the common denominator of all these learning situations is more the word "collaborative" than the word "learning". Still, the variety of uses of the word "learning" reflect two distinct understandings of "collaborative learning": is it a pedagogical method or a psychological process? The pedagogical sense is prescriptive: one asks two or more people to collaborate because it is *expected* that they will thereby learn efficiently. The psychological sense is descriptive: one observes that two or more people have learned and collaboration is viewed as the *mechanism* that *caused* learning. The confusion between the descriptive and prescriptive views leads to frequent overstatements regarding the effectiveness of collaborative learning. I will argue that collaborative learning is neither a mechanism, nor a method:

- Collaborative learning is not one single mechanism: if one talks about "learning from collaboration", one should also talk about "learning from being alone". Individual cognitive systems do not learn because they are individual, but because they perform some activities (reading, building, predicting, etc.) that trigger some learning mechanisms (induction, deduction, compilation, etc.). Similarly, peers do not learn because they are two, but because they perform some activities that trigger specific learning mechanisms. This includes the activities/mechanisms performed individually, since individual cognition is not suppressed in peer interaction. But, in addition, the interaction among subjects generates extra activities (explanation, disagreement, mutual regulation, etc.) that trigger extra cognitive mechanisms (knowledge elicitation, internalisation, reduced cognitive load, etc.). The field of collaborative learning is precisely about these activities and mechanisms. These may occur more frequently in collaborative learning than in individual cognition. However, on one hand, there is no guarantee that those mechanisms occur in any collaborative interactions. On the other hand, they do not occur *only* during collaboration. At some level of description—at least the neurone level—the mechanisms potentially involved in collaborative learning are the same as those potentially involved in individual cognition.

- Collaborative learning is not a method because of the low predictability of specific types of interactions. Basically, collaborative learning takes the form of *instructions* to subjects (e.g. "You have to work together"), a physical *setting* (e.g. "Team mates work on the same table") and other institutional *constraints* (e.g. "Each group member will receive the mark given to the group project"). Hence, the "collaborative" situation is a kind of *social contract*, either between the peers or between the peers and the teacher (then it is a didactic contract). This contract specifies conditions under which some types of interactions *may* occur; there is no guarantee that they *will* occur. For instance, the "collaboration" contract implicitly implies that both learners contribute to the solution, but this is often not the case. Conversely, reciprocal tutoring (Palincsar and Brown, 1984) could be called "a method", because subjects follow a scenario in which they have to perform particular types of interaction at particular times.

In summary, the words "collaborative learning" describe a *situation* in which particular forms of interaction among people are expected to occur, which would trigger learning mechanisms, but there is no guarantee that the expected interactions will actually occur. Hence, a general concern is to develop ways to increase the probability that some types of interaction occur. These ways can be classified in the following four categories; three of them are addressed in this book.

**To Set Up Initial Conditions.** A first way to increase the probability that some types of interaction will occur is to design the situation carefully. The most frequent questions that teachers ask are: What is the optimal group size? Should I select group members with respect to some criteria or leave them to make the group by themselves? Should boys and girls be grouped together? Is it better to have group members who have the same viewpoint or not, the same general level of development or not, the same amount of knowledge with respect to the task at hand or not? Is it better to put them face-to-face or side-by-side? If the interaction is mediated through the network, what are the main features of the groupware to be used? Which tasks are suited for collaborative processes and which are not? These questions have inspired a large body of empirical research on collaborative learning. However, beyond a few main results, it appeared that these conditions interact with each other in a complex way (Dillenbourg et al., 1995); the group heterogeneity effect will, for instance, be different for different tasks. Because of these multiple interactions, it is very difficult to set up initial conditions that guarantee the effectiveness of collaborative learning.

**To Over-Specify the "Collaboration" Contract with a *Scenario* Based on *Roles*.** This approach tends to turn collaborative learning into a method. For instance, reciprocal teaching (see chapter 6) defines a clear specification of roles. Several methods are based on setting up systematic differences among learners, either for triggering conflictual interactions, or else because complementarity of learners" knowledge requires rich interactions: (1) asking subjects to play a specific role in an argumentation, even if the expressed viewpoint is not their personal viewpoint at the outset; (2) giving different visual viewpoints to subjects (Doise, Mugny & Perret-Clermont, 1975); (3) controlling data access in such a way that group members have access to different data ("jigsaw" method). Hoppe and Ploetzner's approach (chapter 8) can be described as a cognitively oriented jigsaw, i.e. based on complementarity between qualitative and quantitative knowledge. This approach defines initial conditions (selecting peers), but it is expected to influence interactions continuously: by selecting problems that cannot be solved solely with one type of knowledge (according to their computational model), the software forces the learners to integrate their respective knowledge.

**To Scaffold Productive Interactions by Encompassing Interaction Rules in the Medium.** The teacher may specify interaction rules for face-to-face collaboration, such as "Everybody in the group should give his or her opinion". In CSCL, interaction rules can be continuously reinforced by encompassing them in the design of the (computer-mediated communication) medium. For instance, various research projects are concerned with "semi-structured" interfaces, i.e. interfaces in which the users communicate with a set of pre-defined buttons. Buttons either form a complete utterance (e.g. "Do you agree?") or open a sentence (e.g. "I propose to . . . .") to be completed with free text. Generally, these buttons characterise the pragmatic value of an utterance (rather than its semantic value) and are hence not surprisingly related to speech acts theory. Interesting results emerged—for instance, that peers focus more on the task and producing fewer off-task comments with semi-structured interfaces (Baker & Lund, 1996; Jermann & Schneider, 1997). However, the case study presented in chapter 3 shows that the interface constitutes a tool—semiotic and physical—that users have to appropriate, and which affects their understanding of the task beyond the simple facilitation/inhibition of particular types of interactions. How strongly or flexibly the interface should shape the interaction processes remains an open issue (Baker, 1997).

**To Monitor and Regulate the Interactions.** The teacher thereby retains a role in the success of collaborative learning. This role is more important as the size of the group increases, and is hence not suprisingly emphasised in chapter 9, where large groups and less-structured tasks are discussed. This role is often named "facilitator" instead of "tutor", because the point is not to provide the right answer or to say which group members are right, but to perform a minimal pedagogical intervention (e.g. provide some hint) in order to redirect the group work in a productive direction or to monitor which members are left out of the interaction. In the context of CSCL, the external regulator needs specific tools for monitoring the interactions that occur in different places and/or at different times. The design of this tool is a main item on the CSCL research agenda. The notion of multiple student models presented in chapter 8 constitutes a first step towards supporting teachers in managing group collaborative interactions. An alternative is to provide the peers themselves with tools for self-regulation of their interaction: for instance, displaying the degree of asymmetry in action or the rate of acknowledgement in interaction (Dillenbourg et al., 1997). We are currently investigating whether peers are able to self-regulate their interaction with this type of feedback.

## 4 The Variety of Meanings for "Collaboration"

So far, I have carefully avoided defining the word "collaboration" itself. However, I am now obliged to address this issue. The adjective "collaborative" concerns four aspects of learning that have been mentioned in the previous section:

1. A *situation* can be characterised as more or less collaborative (e.g. collaboration is more likely to occur between people of a similar status than between a boss and his/her employee, or between a teacher and a pupil).
2. The *interactions* that do take place between the group members can be more or less collaborative (e.g. negotiation has a stronger collaborative flavour than giving instructions).
3. Some learning *mechanisms* are more intrinsicallycollaborative (e.g. grounding has a stronger collaborative flavour than induction), even if, at a very fine level of analysis, learning mechanisms must be similar to those triggered in individual learning (see section 3).
4. The fourth element concerns the effects of collaborative learning, not because this element is used to define collaboration itself, but because the divergent views concerning how to measure the effects of collaborative learning participate in the terminological wilderness of this field.

In this section, I review the criteria applied to each of these four elements for justifying the term "collaborative".

### *4.1 Situations Characterised as "Collaborative"*

Intuitively, a situation is termed "collaborative" if peers are (i) more or less at the same level and can perform the same actions, (ii) have a common goal, and (iii) work together. I will examine these three criteria. The first two concern the degree of symmetry in the interaction. One can discriminate various forms of symmetry, e.g.:

- *Symmetry of action* is the extent to which the same range of actions is allowed to each agent (Dillenbourg & Baker, 1996).
- *Symmetry of knowledge* (or skills or development) is the extent to which agents possess the same level of knowledge (or skills or development). Actually, symmetry is often confused with heterogeneity: two learners may have a similar degree of expertise but different viewpoints of the task.
- *Symmetry of status* is the extent to which agents have a similar status with respect to their community (Ligorio, 1997).

Each type of symmetry can be objective or subjective (Krammer & van Merriënboer, 1989), with interactions being, for instance, affected by the fact than one agent believes his/her partner to be more expert and hence adopts a weaker position in the argumentation. There is no situation of pure knowledge symmetry: there are no two individuals in the world with the same knowledge. Symmetry may change over time: for instance, when partners address sub-tasks for which one peer has particular skills. Chapter 2 points out that the theories relevant to collaborative learning are based on empirical situations with different degrees of symmetry. Research in DAI uses the word "collaboration" for a variety of asymmetrical situations; for instance, hierarchical multi-agent systems. A slight knowledge asymmetry among peers is generally considered as suitable, because it supposedly leads to conflicting interactions. However, despite this interest for knowledge asymmetry, situations depicted as collaborative are generally rather symmetrical, while other situational labels are used for highly asymmetrical situations with respect to actions (control, coordination, etc.) or with respect to expertise (tutoring, teaching, coaching, etc.).

The second criterion is that one generally expects collaborative agents to have *common goals*, whereas competition relies on conflicting goals. However, this criterion is not shared by all researchers, namely those working on sub-symbolic multi-agent systems. During the workshops with the authors of this book, we had the following debate:

> Let us take the case of a robot that moves randomly and drops randomly a piece of wood in a defined space. Another robot, which also moves randomly on the surface, hits the piece of wood and moves it further on. At the end of several thousands of interaction cycles, it may occur that all pieces of wood dragged by all robots will converge towards a common place. This phenomenon is depicted as if the robots collaboratively assembled the wooden pieces (Müller and Rodriguez, 1995). Let us compare this with the case where I lose my car keys, somebody finds them and steals my car. Despite the fact that my distraction did indeed help somebody to steal my car, one would not say that we collaborated on the stealing of my car, because I had no intention of helping the thief to steal it. We would only be viewed as collaborators—and both guilty—if I had deliberately given my key to the thief in order to get money from my insurance company. Similarly, those wood-transporting robots have no common goal (actually, they even have no personal goal); they ignore the fact that their action can create conditions for the actions of another agent. They would be collaborative if a robot computed where it should drop a piece of wood to enable the transportation by a second robot. One robot simply changes the state of the environment, and this change affects the behaviour of another robot.

If one transposes this notion of a common goal from artificial agents to people engaged in real-life situations, one cannot simply assume that partners have completely shared goals, even if some external agent fixes this goal. This is clear, for instance, in the first case study in chapter 9, in which the two nurses and the physician, despite the obvious common goal of getting a new degree, have different expectations with respect to the joint project, because what they learn will affect differently the work practice in their own professional community. Shared goals can only be partially set up at the outset of collaboration; they have themselves to be negotiated (and probably revised as work progresses). Establishing common goals is part of constructing common grounds (chapter 3), since actions cannot be interpreted without referring to (shared) goals and, reciprocally, goal discrepancies are often revealed through disagreement on action. Through the negotiation of goals, agents not only develop shared goals, but they also become *mutually aware of their shared goals.*

The third criterion concerns the degree of *division of labour* among group members. "Collaboration" and "cooperation" are sometimes used as synonymous terms, while other scholars use these terms distinctively according the degree of division of labour. In cooperation, partners split the work, solve sub-tasks individually and then assemble the partial results into the final output. In collaboration, partners do the work "together". However, some spontaneous division may occur even when two people do really work together: for instance, one partner takes responsibility for the low-level aspects of the task while the other focuses on strategic aspects (Miyake, 1986). One could argue that the degree of the division of labour depends on the scale issue: at a high level of granularity, it should always be possible to find some division of labour between the partners. However, when one partner does the task level and the other the meta-communicative level, this leads to a "*horizontal*" division of labour into reasoning layers. This is quite different from the *vertical* division of work into independent sub-tasks as in cooperative situations. The difference is two-fold. First, the layers have to be highly interwoven (one subject monitors the other) while sub-tasks are independent. Second, the horizontal division of labour is unstable in collaboration (it is stable in coaching, for instance), and roles may shift every few minutes with the regulator becoming the regulated, while cooperation refers to a more fixed division of labour that is generally made explicit at the outset. I continue this point in the next sub-section.

These three features (symmetry, shared goals and low division of labour) put chapter 6 into perspective. This chapter considers the cognitive benefits of building and delivering an explanation. However, the methodology used for observing systematically the effects of explaining generally involves experimental situations that are not symmetrical, where the peers do not have common goals and where labour is often systematically divided (explainer/explainee roles). These three differences lead me to think that the effects observed in truly collaborative situations might be very different from those reported in chapter 6, with the results corresponding to a field of research that is still in its infancy.

### *4.2 Interactions Characterised as "Collaborative"*

Another way to define collaboration is to say that it is a situation in which learners interact in a collaborative way. Hence, I review here several criteria for defining collaborative interactions: interactivity, synchronicity and "negotiability".

A first and intuitive criterion is that a collaborative situation should be quite *interactive*. The degree of interactivity among peers is not defined by the frequency of interactions, but by the extent to which these interactions influence the peers' cognitive processes. The degree of interweaving between reasoning and interaction is difficult to define operationally. Hence, I illustrate it with a simple example. Let us assume that two architects must evaluate the value of two houses. If they cooperate, they will each evaluate one house. Their processes are independent: one might count space room-by-room, while the other uses the general dimensions of the house (as a cube). They will interact after a while in order to make their estimations consistent with each other; however, these interactions come after the individual production of partial solutions, and hence will influence them only if these solutions have to be revised. If the two architects interact during the estimation process, one architect might say "Let's count from the basement", thereby leading his partner to measure space room-by-room instead of globally. This example is simple. I consider producing operational criteria for defining the degree of interweaving between interaction and action to be a major challenge for collaborative learning research.

A second intuitive criterion is that "doing something together" implies rather synchronous communication, while cooperation is often associated with asynchronous communication. If agent A has, for instance, to regulate agent B's reasoning, then A and B must work synchronously and hence interact synchronously. This issue became salient in computer-supported cooperative work (CSCW) in which communication tools are often characterised as synchronous or not (see chapter 9). In fact, this dichotomy corresponds to the underlying technology but not to the real performance of communicative systems. For instance, while electronic mail is classified as asynchronous communication, chat-box software is classified as synchronous. If a user in Geneva communicates with a colleague in London, the delivery delay might be 2 seconds in a chat box and 20 seconds by electronic mail. Where is the threshold beyond which one considers communication to be asynchronous? We controlled the delay of acknowledgement in a MUD environment, i.e. a text-based virtual reality including typed synchronous interaction (Dillenbourg & Traum, 1996). The average delay was 48 seconds, i.e. longer than by e-mail. This long delay may partially be due to lack of attention (the receiver is working on something else), but it includes the time necessary to produce the answer, i.e. to reason about it and to type it. Users of MUDs and other groupware systems have developed conversational rules for coping with this delay, to keep the floor, to initiate turn taking, and so forth (O'Conaill, Whittaker & Wilbur, 1993). In other words, synchronicity is less a technical parameter than a social rule. It is a *considerate meta-communicative contract*: the speaker expects that the listener will wait for his message and will process the message as soon as it is delivered. If the medium breaks the conversational rules established for another medium, users create new ways of maintaining this *subjective feeling of synchronicity of reasoning*. In these same MUD experiments, we observed, for instance, that this synchronicity of reasoning often leads partners to make the same utterance simultaneously. One may hypothesise that the feeling of reasoning synchronously increases the process of mutual modelling, i.e. the effort produced by one agent to model the knowledge state of his partner.

Another feature of collaborative interactions is that they are *negotiable*. A main difference between collaborative interactions and an hierarchical situation is that one partner will not impose his view on the sole basis of his authority, but will—to some extent—argue for his standpoint, justify, negotiate, and attempt to convince. Hence, the structure of collaborative dialogue is expected to be more complex than, for instance, tutoring dialogues. This point leads to several comments with respect to the chapters of this book:

- The degree of negotiability may be a difference between an explanation, as studied in formal experimental settings reviewed in chapter 6, and the explanation that emerges from naturalistic collaborative situations.
- In addition to negotiating at the task level, collaborative partners usually have the possibility to negotiate how to interact (meta-communication): for instance, in order to clarify the pragmatic value of an utterance (e.g. "Is that a question or a claim?") or to tune turn-taking ("Let me speak"). Meta-communicative mechanisms have been identified in chapter 7 as an important difference between the operators used by Michalski in machine learning and those by Baker in computational models of dialogue.
- Negotiation can occur only if there is *space for negotiation* (Dillenbourg & Baker, 1996), i.e. if something can actually be negotiated. Negotiation at the meta-communicative level can be inhibited by forcing partners to play well-defined roles (a growing trend in research on collaborative learning). Negotiation at the task level can be inhibited in trivial tasks, in which there is nothing to disagree upon, and in tasks with semantically obvious and unambiguous terminology, in which there is nothing to misunderstand. The boundary between misunderstanding and disagreement is shallow. If we do not understand each other, one cannot say that we properly agreed. The grounding process is the bottom layer of the negotiation process (Dillenbourg & Traum, 1996). This grounding + negotiation process, i.e. the way through which partners can build a shared solution, hence became a central concern for research in collaborative learning. Chapter 3 is an ambitious attempt to relate a linguistic theory, developed by Clark and several colleagues, with the learning mechanisms described by the cultural–historical psychology. This chapter reflects a general trend: the growing contribution of linguistics—especially pragmatics—to theories of collaborative learning.
- In grounding, the *space for misunderstanding* plays the same role as the space for negotiation in agreement. It constitutes an important element in modelling the dynamics of collaborative learning. Of course, any communicative act offers opportunities for misunderstanding. However, this space is, for instance, more narrow when partners negotiate the next action to be performed, or the meaning of an utterance (as in chapter 3) versus the situations in which the partners negotiate which problem they have to solve (as in chapter 9). Chapter 4 draws attention to the fact that the use of ambiguity-free (or noise-free) communication protocols in multi-agent systems, despite being understandably related with effectiveness matters, might be detrimental to the modelling of complex collaborative processes. Similarly, the linguistic model that underlies chapter 3 uses the concept of *least*

*collaborative effort* to reach mutual understanding (Clark & Brennan, 1991), referring to the fact that the cost of producing an ambiguity-free utterance may be higher than the cost of collaboratively repairing the misunderstandings that may arise. The study of collaborative learning involves a different viewpoint: when two partners misunderstand, they have to build explanations, justify themselves, reformulate statements, and so on—all of these being activities that can lead to learning. Hence, in contrast with DAI and linguistic criteria, a collaborative learning situation should not eliminate (if possible) the probability of misunderstanding (in order to reduce the cost of communication), but instead leave the space for misunderstanding that is necessary for sustaining the subjects' efforts to overcome miscommunication. Of course, a too high level of misunderstanding would be detrimental to learning. We proposed the notion of *optimal collaborative effort* (Dillenbourg & Traum, 1996) to indicate that grounding efforts have to remain subordinated to the accomplishment of the task and the production of rich interactions.

### *4.3 Processes Characterised as "Collaborative"*

In the previous discussion of synchronicity, I started by discussing synchronous interactions and moved to a discussion of synchronous reasoning and mutual modelling. In other words, the topic moved from defining collaborative interactions to defining mechanisms that are specific to collaborative learning. I now explore this direction. However, I said earlier that, at the most fundamental level, the learning mechanisms involved must be those that operate in the case of individual cognition, since there are still individuals agents involved in group interactions. In this section, I review the mechanisms that have been studied with individuals and then extended to pairs, and the mechanisms that seem to be more specific to collaborative learning. I start with mechanisms that are known to be central to individual cognition:

- *Induction.* Schwartz (1995) observed that pairs draw more abstract representations of the problem at hand, because their joint drawing had to integrate what was common to the representations built by each individual. The basic underlying process is inductive, i.e. to keep as relevant the features that are present in both representations. I would even argue that the grounding mechanism is inductive since the learner must induce patterns that relate referring expressions with referents.
- *Cognitive load.* Another concept borrowed from individual cognition is that of cognitive load. In collaboration, the horizontal division of labour into, for instance, task-level and strategy-level tasks reduces the amount of processing performed by each individual. Reduced cognitive load may explain why regulating one's partner is easier than self-regulation and therefore why group members improve their regulatory skills (Blaye, 1988). Conversely, the interaction with other group members increases the cognitive load, which is not detrimental in itself (e.g. building an explanation is a cognitively demanding task, but still it leads to

learning), but may become detrimental in case of over-load. For instance, Schnotz (1997) observed that while individuals perform better with animated pictures than with static pictures, pairs perform better with static pictures. He interprets this difference by the cumulative effect of the cognitive load involved in parsing dynamic pictures and the cognitive load due to social interaction. Chapter 4 investigates this point in the case of artificial agents. The balance between the reduction of individual computing due to division of labour and the increase of individual computing necessary for interaction (or central coordination) has been systematically investigated in DAI (Durfee, Lesser & Corkill, 1989; Gasser & Huhns, 1989). The cost/benefit ratio of splitting the work is a main criterion for comparing multi-agent architectures.

- *(Self-)explanation.* The case of explanation is different from the previous examples. The concept of explanation is, in itself, related to social situations, but it has been imported into studies of individual cognition, both in psychology (the "self-explanation effect") and in artificial intelligence (explanation-based learning). The issue addressed in chapter 6 is whether the effects established with "explanation for solo agents" can now be re-exported to collaborative settings, and what needs to be added to or removed from existing solo models to fit with real (social) explanations.
- *Conflict.* The concept of conflict also concerns both the intra-individual and inter-individual planes. Although it is intrinsically a social concept, it was developped in cognitive science in the Piagetian theory, i.e. individual-centred theory of development. Piaget's co-workers re-exported this concept to the case of social interactions, where a discrepancy between the knowledge or viewpoints of two peers leads to conflicting statements or positions with respect to the task at hand (Doise & Mugny, 1984).

The processes listed above are not specific to collaborative situations, they simply may occur more often or more spontaneously in those situations. Are there other learning processes that would be very specific to social interactions?

- A short answer is the *internalisation* process, i.e. the "transfer" of tools from the social plane (interaction with others) to the inner plane (reasoning), since this process implies social interaction. Actually, this process has been more studied in asymmetrical situations (adult–child interaction being viewed in the Vygostkian approach as the "motor" of development) rather than in collaborative settings as such (see chapter 2). Unfortunately, internalisation describes more an effect than its underlying mechanisms: one can observe, for instance, that some concept is first introduced by an adult for solving the task with a child and then later on used by the child to reason about the task, but the successive steps that lead the child to "think with" the concept are not clear. Wertsch (1985), however, gave some indications regarding the process itself. He showed that the main cognitive change was less the transition from the inter- to the intrapsychological plane, but, within the interpsychological phase, when the child becomes able to use the to-be-internalised concepts in his/her conversation with his/her mother.

- The authors of chapter 3 might rephrase Wertsch's findings by saying that the concept has been grounded, i.e. became part of the common grounds built by the mother and her child. Another often quoted mechanism is "*appropriation*" (Rogoff, 1990), by which an agent re-interprets his/her own action or utterance under the light of what his/her partner does or says next (Fox, 1987). Again, the similitude with the grounding theory is striking. These similarities exist at two levels. At the interaction level, a rephrasing of what has already been said by one's partner would be viewed as repairing a misunderstanding in the grounding theory and as reformulating the previous utterance within a more elaborated conceptual framework for the socio-cultural theory. Chapter 3 investigates these similarities. At the cognitive level, both mechanisms assume that each partner maintains some representation of the other, and that each partner infers from his/her partner's actions or utterances what he/she is thinking, what he/she believes, and so on.
- This "model of the partner" may not be very detailed and accurate, not updated permanently, and will, of course, be of a lower quality as the number of group members increases or as the frequency of interaction decreases. It is certainly not complete, in the sense that it will be developed only up to the point where it needs to be developed to succeed in performing the task at hand (such as the grounding criterion, which is also based on its functionality). Research on multi-agent learning also pays attention to this process (see chapter 4). Further research needs to be carried out in order to determine (1) how elaborated is the mutual representation that partners build; (2) which processes are involved in mutual modelling; and (3) the extent to which these processes generate cognitive effects at the task/knowledge level. An hypothesis mentioned in chapter 7 is that *mutual modelling* implies some differential reasoning (comparing what I do or know with what my partner does or knows) and that the perception of discrepancies with respect to one's partner's knowledge increases the awareness of one's own knowledge.

### *4.4 Effects of Collaborative Learning*

Most research on collaborative learning has attempted to measure its effects, generally through some individual pre-test/post-test gain with respect to task performance. Some more-specific effects have been described in terms of conceptual change (Amigues, 1987; Roschelle, 1992; Pea, 1993) or increased self-regulation (Blaye, 1988; Gilly, 1989). The choice of these dependent variables leads to two methodological issues.

The first issue could be stated as follows: "**effects of what?**". As it appears from the above sections, a collaborative learning situation includes a variety of contexts and interactions. Talking about the effect of such a broadly defined term would be as meaningless as talking about the benefit of taking a medicine, without specifying which one. One should not talk about the effects of collaborative learning in general, but more specifically about the effects of particular categories of interactions (Dillenbourg et al., 1995). This implies controlling *a priori* which types of interactions will occur, as in some experiments on explanation reported in chapter 6, or analysing *a posteriori* which interactions did actually take place during collaboration, as in dialogue studies reported in chapter 3. The latter studies can be criticised for being too qualitative, leading to few conclusions, while the former can be attacked for their poor ecological validity. Both studies reflect, however, an important trend: researchers no longer (or should not) treat collaboration as a "black box", but **zoom in** the collaborative interactions in order to gain better understanding of the underlying mechanisms. In this context, CSCL environments are a very interesting tool since they enable both a detailed recording of all interactions and the careful design of the empirical situation.

The second methodological issue concerns the mode of evaluation. The effects of collaborative learning are often assessed by **individual task performance measures**. It has been objected that a more valid assessment would be to measure group performance. This "validity" can be understood in practical terms: more and more professionals have to collaborate and it is an important goal for any educational institution to improve the students' performance in collaborative situations. The validity issue has also a theoretical side: from the distributed cognition viewpoint, assessing groupwork through individual performance would be as meaningless as assessing a painter without his brush (Perkins, 1993). Within the group evaluation approach, one may verify whether the performance of a specific group has increased or assess if group members have developed some generic ability to collaborate—which they could then re-use in other groups. However, the existence of this hypothetical ability to collaborate, although intuitively shared by many staff recruiters, remains to be established, at least in cognitive science.

## 5 Theories of Collaborative Learning

I have often used the following definition of collaboration:

> ". . . a coordinated, synchronous activity that is the result of a continued attempt to construct and maintain a shared conception of a problem" (Roschelle & Teasley, 1995, p. 70)

Compared with the four criteria listed in the previous section (situation, interactions, processes and effects), the definition of Roschelle and Teasley does not include the situation. A mutual effort of shared understanding does also occur in non-collaborative situations: for instance, in teacher–pupil interactions (Douglas, 1991), and in virtually all verbal interactions. However, this definition constrains the three other items. Shared understanding can be viewed as an effect, if the goal is that a group builds the common grounds necessary to perform well together in the future. It can be viewed as a process by which peers perform conceptual change. It can be viewed as a condition for conducting effective verbal interactions.

A theory of collaborative learning concerns these four items: criteria for defining the *situation* (e.g. symmetry, degree of division of labour), the *interactions* (e.g.

symmetry, negotiability), *processes* (e.g. rounding, mutual modelling) and *effects*. The main theories reviewed in chapter 2 (socio-cognitive, socio-cultural, etc.) cover the four aspects, while more-local theoretical contributions (e.g. the self-explanation effect) cover a specific aspect. The key for understanding collaborative learning is in the relations between those four items. At a first glance, the situation generates interaction patterns; these interactions trigger cognitive mechanisms, which, in turn, generate cognitive effects. However, such a linear causality is a simplification. Most relations are reciprocal.

- There is a bi-directional link between the situation and the interactions: on one hand, the situation defines the conditions in which some interactions are likely to occur, but, on the other hand, some situations are labelled "collaborative" because the interactions that did occur between members were collaborative.
- There is a bi-directional link between the interactions and the processes, as illustrated above by the relationship between synchronicity and mutual modelling: I needed to refer to cognitive process (mutual modelling) in order to define a feature of interaction (synchronicity) beyond simple technical terms.
- There is a bi-directional link between the processes and the effects of collaborative learning. In principle, the processes generate the effects. However, some processes are described by the effects, such as "internalisation". Conversely, some effects are expressed in terms of group processes, such as the ability to work in a group. This ambiguity is not specific to the field of collaborative learning: for instance, induction can be viewed by a psychologist as a process while it might be viewed by a neuro-physiologist as the output of complex chemical processes.
- This ambiguity also concerns the distributed cognition theory, where the very idea of viewing the group as a unit can be understood both as a theoretical standpoint (group interactions being described as cognitive processes) or as the description of an effect or an achievement (some groups interact so well that they succeed in forming a distributed cognitive system).

## 6 Multidisciplinarity in Studies of Collaborative Learning

This book is intended to illustrate the benefits of collaboration between scientists from psychology and computer science, namely machine learning. Most chapters have been co-authored by scholars from both sides. The coordination of these views is especially important for chapters 4, 5 and 7. Two remarks must be made, however. First, machine learning has not devoted a lot of attention to collaborative learning. The idea of collaborative problem solving has been explored in DAI, where the focus was on immediate performance. It is only recently that more attention was paid to learning processes in multi-agent systems (Weiss, 1997). Moreover, it is not unfair to say that DAI has developed independently from the cognitive theories reported in chapter 2 and probably from linguistic theories. Chapter 4 describes some differences between multi-agent systems and cognitive psychology. Chapter 5 builds the comparison on a concrete case study, where robots and humans have to perform the same task collaboratively. Chapter 7 compares the basic operators used to describe processes involved in machine learning and dialogue studies. In some way, these three chapters are rather modest in their scope and in their results. Their modesty has, however, to be compared with the width of the gap that separates cognitive psychology and DAI.

However, a few elements have helped to reduce the gap. First, this book focuses on cognitive psychology and hence deliberately neglects the socio-affective aspects of collaborative learning (it is also one of its weaknesses). This choice does not imply that social and affective aspects are not crucial to understanding human–human collaboration, but that the cognitive layer is probably the easiest to start with when engaging in a dialogue with computer scientists. Second, the dialogue was made easier by the fact the psychologists involved in this book all had a computer science experience, either because they were involved in computational models of cognition (e.g. chapters 6 and 8), or because they were involved in CSCL (chapters 3, 7, 8 and 9). Computer-supported collaborative learning is a very promising area for research, not only because it answers to a strong social demand, but also because it enables scholars to "zoom in" collaboration. By very carefully designing the situation, tuning up the interactions supported, and by recording systematically all interactions, CSCL is the "microscope" for scientists in this field. It also introduces a continuum between DAI and psychology, and an intermediate area where research is intrinsically multi-disciplinary.

# 2

# Learning Together: Understanding the Processes of Computer-Based Collaborative Learning

Karen Littleton and Päivi Häkkinen

## 1 Introduction

Whilst the study of collaborative learning has a relatively brief history, there have nevertheless been substantial changes in the nature of the research being undertaken in this field. Initially, the primary aim was to determine whether and when collaborative learning was more effective than learning alone (Dillenbourg et al., 1995) and there is a substantial body of empirical evidence demonstrating the positive effects of social interaction for learning (e.g. Tudge & Rogoff, 1989; Teasley & Roschelle, 1993; Light et al., 1994). As the results of these studies emerged, it became apparent that computer use can constitute a particularly valuable context for social interaction (Crook, 1996). As a result, the need to design learning environments that facilitate social interaction, cooperation and collaboration in the classroom has increasingly been recognised (Vosniadou, 1996). Recently, however, interest has shifted away from considering just the outcomes and products of collaborative work, towards analysing interactions as a means of gaining insight into the processes of collaborative learning. The aim of such analyses is to identify what constitutes productive collaborative activity. In the sections that follow, we outline some of the theoretical approaches currently informing the analysis of learning interactions and demonstrate how the study of collaborative activity in computer-based settings can enhance our understanding of the processes of productive interaction. In doing so, we focus primarily on research studies of collaborative educational computing activity conducted predominantly in school settings.

## 2 What is Collaborative Learning?

To a lay-person, the term "collaboration" is unproblematic—collaboration simply refers to the fact that people are working together on a task. In the research literature, however, much has been written about how best to define collaboration and collaborative learning. Even a cursory review of the recent literature on learning and instruction reveals a wide range of terminology relating to collaborative learning (see also chapter 1, this volume). Terms such as "peer collaboration" (Tudge & Rogoff, 1989), "collaborative learning" (Forman & Cazden, 1985), "coordinated learning" (Koschmann, 1994) and "collective learning" (Pea, 1994) appear frequently. Each term carries with it its own particular nuances and emphases. Thus, for example, Pea (1994, p. 286) argues for his choice on the grounds that: "not all learning feels or probably is collaborative; it is sometimes competitive or coercive in nature".

Whilst the profusion of terminology can undoubtedly result in a degree of confusion, the breadth of associated meanings do successfully characterise the diverse and varied nature of interactions that occur between learners. Moreover, there is consensus amongst researchers that collaboration involves the construction of meaning through interaction with others and can be characterised by a joint commitment to a shared goal. The importance of a common goal is central to many accounts of collaborative activity, including analyses of people working together in emergency situations (Barfurth, 1995).

Whilst the terms collaborative and cooperative learning are often used interchangeably, collaborative activity requires more than the effective division of labour that constitutes cooperative work. Collaboration necessitates that participants are engaged in a coordinated effort to solve a problem or perform a task together (Teasley & Roschelle, 1993). One can go further still, distinguishing collaborative learning from other forms of peer interaction by advocating that collaborative activity involves the construction of a solution that otherwise could not be produced (Forman & Cazden, 1985; Tudge & Rogoff, 1989). More recently, the need for reciprocal understanding between collaborators has been highlighted as an essential prerequisite for collaborative learning. Such reciprocity involves individuals' establishing, through the negotiation of meaning (Nystrand, 1986), mutually shared or "common" knowledge; the assertion being that coordinated cognitive activity depends upon establishing intersubjectivity (Rogoff, 1990; Wertsch, 1991).

Underpinning these definitions of collaborative learning are detailed studies of learners working together to learn and to solve problems. In the sections that follow, we draw on examples of work conducted within different traditions to illustrate some of the ways in which collaborative interactions have been studied and conceptualised within the psychological and educational research literature.

## 3 Socio-cognitive Conflict

The potentially productive role of conflicting perspectives on the task in hand was first highlighted within the work of Piaget, and assumed prominence within the neo-Piagetian tradition of research. Whilst much of this work ante-dated the advent of computers, this tradition has nevertheless exerted considerable influence on contemporary research investigating productive interactions around computers.

In general, Piagetian theory does not provide an adequate conceptualisation of the role of social interchanges in stimulating and shaping intellectual development. Social development is regarded simply as proceeding in parallel with cognitive development, and sharing the same underlying organisational principles. Given that Piaget believed that children's active construction of their own understanding is fundamental to their cognitive growth, it is perhaps not surprising that he was deeply opposed to transmission from adult to child as a model for cognitive development. Interaction with adults was seen at best as irrelevant, or at worst as detrimental, interfering with children's exploration of their physical environment and hence the construction of their understanding. By contrast, however, peer interaction represents a peculiarly potent source of progress (Piaget, 1932).

Whereas, according to Piaget, the young child cannot treat adults' ideas "on their own merit" because of the differences in power and status involved, disagreements with other children serve to highlight alternatives to the child's own particular point of view. Since the alternatives can be considered "on equal terms" and the resulting conflicts of opinion necessitate resolution, the children involved can be prompted towards higher-level solutions, which incorporate the partial insights reflected in their initial positions. In essence, then, the germs of intellectual progress were seen in the conflict of perspectives.

This account of peer facilitation of children's understanding is primarily linked to Piaget's account of the genesis of "operational thinking". Nevertheless, through the work of Doise and colleagues, it has come to exert a substantial and wide-ranging influence. During the 1970s and 1980s, Doise, Mugny and Perret-Clermont (e.g. Perret-Clermont, 1980; Doise & Mugny, 1984) conducted a series of studies investigating whether *individual* progress in understanding could be promoted by "socio-cognitive conflict"; i.e. exposure to the conflicting ideas of peers in the context of paired or small-group problem solving. These Genevan researchers were not simply interested in whether groups of children working together would demonstrate better on-task performance than individuals; rather, the key research question was whether children who were given the opportunity to work on a task together would make greater individual progress than those who were not given such an opportunity. In order to answer this question, Doise and colleagues designed studies that comprised individual pre-test and post-tests sessions, separated by a period of work on a task undertaken either individually (in the "control" condition) or in pairs/groups (in the "experimental condition").

Where greater individual gains were evident in the experimental group, the associated explanation was couched in terms of socio-cognitive conflict. The resolution of such conflict, however, was conceived of in individualistic terms. Such resolution did not necessarily occur during the session itself, but might do so later. Moreover, Doise was able to demonstrate that simply attributing conflicting views to another child, even though that child was not physically present, was sufficient to bring about progress. So, in terms of understanding the productivity of collaborative interactions, according to this account a social process of sorts is involved, but the role of interaction *per se* is minimal.

The three-step (individual pre-test/paired vs. individual treatment/individual post-test) design utilised by Doise and colleagues has been widely adopted by researchers investigating peer facilitation of computer-based learning, as has the use of relatively brief and circumscribed tasks. These features, taken together with the rather self-contained nature of much computer-based interaction and the convenience of computer-based data collection, afford ideal conditions for experimental methods. The individualistic, cognitive level of analysis thus came to be associated with the adoption of experimental methods of research in this field.

An example of this level of analysis—the use of the experimental method and an interest in the potentially productive role of conflict—can be found in the work of Light and Glachan (1985), who studied children working on a computer version of the pegboard code-breaking game "Mastermind". The computer generated a digital sequence of code that the children had to "crack" by trying out sequences and amending their responses in light of subsequent feedback. Those children who worked together on the code-breaking task enjoyed some advantage over those who worked alone. In an attempt to understand the basis of this facilitation effect, Light and Glachan tape-recorded the children's interaction. They then analysed the taped interaction sessions, looking particularly for evidence of "arguments". Arguments were defined as sequences where the children were offering different suggestions as to what move to make next, and where at least one of them was offering some form of justification. Light and Glachan's analyses revealed that arguments were common in about half of the pairs and scarce in the other half. Significantly, the improvement of children's scores from pre-test to post-test was greater for those pairs who engaged in frequent argument than it was for the remainder. These findings offer some support for the view that socio-cognitive conflict, or at least disagreement, may represent a valuable condition for progress in peer interaction situations. Further, more recent evidence regarding the role of conflict also emerges from a series of experiments undertaken by Howe and colleagues. The design of these studies has followed the familiar pattern of individual pre-tests and post-tests, and in between them a session in which children work together in pairs or small groups. The participants, however, were older and were using computer-based tasks related to the physics curriculum.

In a study concerning the trajectories of objects dropped from planes, Howe, Tolmie and Anderson (1991) formed pairs on the basis of pre-test responses such that students in a pair were either similar or different in terms of their initial predictions and conceptions. Those pairs showing the greatest pre- to post-test gains were those where the pair members differed both in terms of their initial predictions and in the conceptions on which these predictions were based. In a subsequent study (Howe et al., 1992), the researchers differentiated between the actual judgements that learners made and the strategies they used in approaching the problem. In this study, the problem involved making a comparison of the speed of two trains under various conditions. One of the trains ran across the computer screen at a constant speed, whilst the other train either accelerated or decelerated, with variations of direction, starting position and so on. Pre-tests were used to establish the judgements made and strategies used by each of the participants, and then the pairs were composed so that the partners were either similar or different in judgements or strategies. The most productive group combination in terms of conceptual change was where partners differed in both strategy and judgement.

Whilst the findings described above offer persuasive evidence to suggest that socio-cognitive conflict may be a key ingredient in peer facilitation of learning, other researchers have raised doubts about the pivotal role of conflict and have criticised the concept as vague, ill-defined and hard to operationalise outside "experimental" research settings (e.g. Blaye, 1988). Some have also pointed to evidence which suggests that, in certain circumstances, peer interaction can result in both regression and development (Tudge, 1989), and that interaction between children at the same level is actually sometimes the most effective situation (Light et al., 1994). So, under certain circumstances, collaborative gains may have little to do with decentring through conflict in Doise's sense, but may have more to do with the socially mediated processes of conflict resolution. Concepts such as "argumentation" and "negotiation" take us towards a much more fundamentally social model of productive interaction.

## 4 The Joint Construction of Understanding

Piaget emphasised social interaction, and more specifically peer interaction, from the perspective of its specific role in the development of logical reasoning. Vygotsky (1978), however, conceptualised social interaction as being at the core of the developmental process. His approach therefore provides an altogether more central and constructive role for adults in promoting children's development, and specific emphasis is placed on the importance of "asymmetrical" interactions between partners who differ markedly in their levels of ability. Seen from a Vygotskian perspective, the developmental process is a constructive one, but it is not the child alone who is doing the constructing (Light & Woodhead, 1991). Rather than being solely based on experience with the physical world, the construction of knowledge and understanding is seen as a fundamentally social activity. In and through this activity, the cognitive resources of the whole society are gradually made available to the child as she/he participates in practices embodied in the cultural life of the community, artefacts, technologies and rituals (Crook, 1996). Individual cognitive development is therefore seen as a process of internalising socially regulated and mediated knowledge. Processes that go on between the child and others become the basis for processes that subsequently go on within the child—discussion, interaction and argument become internalised as the basis for reflection and logical reasoning (Light & Woodhead, 1991). In this way, the child's understanding of the world is mediated by and built up through interaction with others, and meanings are negotiated and established through interaction in a wide range of social contexts.

This "social constructivist" approach, with its emphasis on negotiation of meaning and processes of joint construction of understanding, has been adapted by a number of contemporary researchers to embrace the situation of children of more or less the same level of ability working on a task collaboratively (e.g. Forman & McPhail, 1993, as cited in Forman & Larreamendy-Joerns, 1995). Here we are invited to consider the role of collaborative rather than conflictual processes, and particularly the constructive role of discussion. The group, rather than the individual, becomes the focus of analysis, and the emphasis is placed on socially constructed properties of interaction, rather than on individual performance on pre- and post-tests.

Mercer (1995) offers an analysis largely in this vein. Inspired by a theory of language and cognition that is essentially socio-cultural, Mercer and colleagues focused on the nature of primary school children's talk when they were working together to solve problems at the computer in the context of the classroom. Three qualitatively different types of talk were identified—disputational, cumulative and exploratory talk. Disputational talk is unproductive disagreement and can be characterised by an initiation (e.g. proposition, hypothesis, instruction) followed by a challenge (be this a direct rejection or a counter-proposition/hypothesis). Such challenges typically lack clear resolution or else result in resolution that is not supported by agreement. Cumulative talk simply adds uncritically to what has gone before. Initiations are typically accepted either without discussion or with only superficial amendments. In contrast, however, exploratory talk demonstrates the active joint engagement of the children with one another's ideas. Whilst initiations may be challenged and counter-challenged, appropriate justifications are articulated and alternative hypotheses offered. Where alternative accounts are offered, they are developments of the initiation. Progress thus emerges from the joint acceptance of suggestions. Whilst Mercer and colleagues acknowledge that all three types of talk are appropriate in certain circumstances, they maintain that exploratory talk offers a potential for learning over and above that offered by the other types of talk. Exploratory talk is characterised as the embodiment of critical thinking—essential for successful participation in educated communities of discourse. It is argued that this type of talk represents a distinctive social mode of thinking and that, by encouraging the awareness and use of such talk, we may be able to help learners to develop intellectual habits that will serve them well across a diverse range of situations (Mercer and Wegerif, in press). Seen in these terms, supporting certain types of classroom discourse enables meaningful, socially organised learning to occur.

As with much of the research on peer interaction conducted from a neo-Vygotskian perspective, the work of Mercer and colleagues has been classroom-based, and has involved the extended and naturalistic observation of groups of children. Such extended and naturalistic classroom-based studies stand in striking contrast to the experimental research reviewed earlier in this chapter, which typically requires the use of very self-contained "experimental tasks", involving children working for one or more sessions on a purpose-made or "off-the-shelf" piece of software that involves a singular session of collaboration in which the partners are enjoined to work together intensively at the computer. It is not clear that the analysis of such activity offers an adequate model of children's day-to-day learning environment, in which learning, whether with the computer or otherwise, is typically a more continuous and extended process. Naturalistic classroom-based studies, such as those undertaken by Mercer and colleagues, potentially afford opportunities to reflect on the extended aspects of collaborative learning using computers. Understanding the creation of shared knowledge over extended periods of time will bring with it many challenges. For instance, as learners create a shared history and develop common knowledge, the need to be verbally explicit about their work will decline. For the participants in a collaboration, this is undoubtedly an asset. For researchers of collaborative interaction, who have typically studied forms of verbally explicit reasoning, this poses a significant challenge. How best to study that which is tacit and unspoken?

The emphasis on negotiation and the joint construction of understanding, so central to the neo-Vygotskian analyses of Mercer and colleagues, finds echoes in work that possesses other, diverse, theoretical underpinnings. Consider, for example, the analyses of collaborative activity undertaken by Teasley and Roschelle (1993). They investigated the collaborative construction of new problem-solving knowledge with specific reference to individuals' coordinated effort to negotiate and share meanings. These researchers argue that collaborative problem solving enables the construction of a shared conceptual structure, which they call a joint problem space (JPS). A JPS integrates semantic interpretations of goals, features, operators and methods, and students participate in its creation and maintenance.

As a means of studying the creation and maintenance of a JPS, Teasley and Roschelle analysed the activity of collaborative dyads engaged in problem-solving activity using a computer simulation, called the Envisioning Machine, which portrays a graphical, dynamic simulation of a physicists' mental model of velocity and acceleration. Informing their analyses were the influences of several theories and the methods of cognitive science, conversation analysis, activity theory and micro-analytic research (Schoenfeld, 1992).

Teasley and Roschelle's analyses revealed that students use the structure of conversation to continually build, monitor and repair a JPS. The turn-taking structure of "collaborative completion", in which one partner's turn begins an idea and the other one uses the next turn to complete it, proved particularly useful in understanding the process of shared problem solving. In conflict situations, the dyads had a tendency to maintain a shared conception of the problem by resolving misunderstanding, and they used "repairs" to deal with problems in speaking, hearing or the comprehension of dialogue. The use of "narrations" enabled partners to monitor each other's actions and interpretations. However, language was not the only tool used to maintain shared understanding—the use of action and gesture were crucial too. Such findings highlight the importance that non-verbal interaction for collaborative learning needs to be investigated more systematically.

## 5 Collaboration, Classrooms and Community

Many of the analyses presented thus far are, perhaps as a result of their focus on microgenetic processes, in danger of decoupling educational computer activity from the broader context of classroom life. Thus, there is a need for collaborative educational computing activity to be studied in ways that do not isolate it from the wider "cultural niche" that is the classroom.

To date, much of the work on collaborative learning has focused on understanding collaborations between dyads, or small groups, working relatively autonomously at the computer. However, if we are fully to understand the processes of productive interaction, we need to recognise the powerful influence of the teacher on children's computer-based collaborative work. It is the teacher's responsibility to ensure that children's computer-based experiences contribute to their education and, as Mercer and Fisher (1992) note, this is a responsibility that cannot be delegated to even the most sophisticated software or to the children themselves. There is, therefore, a need for rich descriptions, conceptualisations and evaluations of the ways in which teachers attempt to "scaffold" children's collaborative learning with computers and how this, in turn, impacts on the processes and outcomes of their joint work. The importance of understanding such everyday interactions amongst members of learning communities is emphasised by theorists working within the field of "cultural psychology".

As we saw in the previous section, Vygotsky's work initially provided a starting point for an approach to understanding how peer interaction can facilitate learning and problem solving. This approach highlights the joint construction of solutions to problems, with solutions being achieved predominantly through discussion (Mercer, 1995). Vygotsky's work has, however, also provided an important starting point for a wider theoretical development known as "cultural psychology" (Crook, 1996). Cultural psychology focuses on thinking and reasoning as activities that take place in particular situations. This approach has come together with influences from anthropology (Lave & Wenger, 1991) and cognitive science (Suchman, 1987) to form the basis of what is now commonly termed "situated learning".

Situated approaches to understanding human cognition and learning encompass many diverse theoretical perspectives. Yet, underpinning all these perspectives is a recognition that human cognition and learning are embedded in specific contexts (e.g. Brown, Collins & Duguid, 1989) and are constituted through processes of interdependence, interaction and transaction between people and their environment (e.g. Sternberg & Wagner, 1994). This view of cognition as socially and ecologically grounded has certain implications for our theorising on the nature of collaborative interaction. We need, for example, to recognise that the study of interactions conducted in educational settings and based around predominantly school-related tasks offer us a highly context-specific understanding of the process of collaborative work and interaction. In classroom settings, teachers and pupils engage in heavily contextualised discourse. Their interactions are influenced by a complex contextual system that is inseparable from how education is defined in our culture. Interactions are thus framed by, and therefore can only ever be fully understood within, the context of particular institutional settings. This is because different institutional contexts afford, and indeed constrain, different opportunities for interaction.

"Situated" approaches to understanding cognitive activity also stress the notion of "mediation" and the role of cognitive tools and artefacts. The concept of mediation refers to the fact that our relationship with the outside world is always mediated by signs and artefacts (Säljö, 1995). Put very simply, we think with and through artefacts. The computer is an artefact that is not only capable of supporting collaborative endeavour, but also has the potential to transform uniquely the social processes of children's joint problem solving. Yet, in many studies of collaborative work, the central role of the computer in framing and mediating joint activity is seldom noted, let alone conceptualised. Findings from a study by Järvelä (1995), in which the learning interactions were organised according to the principles of a cognitive apprenticeship model (Collins, Brown & Newman, 1989), demonstrates the value of studying the unique role of the computer in mediating paired work. Järvelä applied a systemic approach to studying classroom learning in secondary school students who were engaged in the construction of Lego models of automated machines. The functioning of these machines was controlled by LOGO programs. The computer instructional technology played a vital role in supporting the student's collaboration as the cognitive work of the students was externalised and made jointly available in the interaction. The concrete object of the problem to be solved increased the cognitive on-task talk amongst peers, enhanced meta-cognitive activity and facilitated scaffolding discussions between teachers and students.

The ways in which the computer can serve to re-organise the social processes of collaborative work have also been recognised by Keogh and Barnes (1994). They studied the nature of the talk and joint activity in same- or mixed-gender pairs of children engaged in an identical language problem-solving task (involving the assembly of a poem from a jumbled collection of phrases) presented either on or off the computer. Their findings illustrate the impact that the presence of the computer has on the activity of the mixed-gender pairs: boys dominated the task and decision making when it was presented on the machine, whereas the activity was distributed equally between the pair members when the task was off-computer (Säljö, 1995).

However, as well as re-organising interactions between learners, computer technology can furnish new educational environments for collaboration. Consider, for example, the opportunities afforded by computer-mediated communication. Here, computer technology can support joint activity that needs to be separated in time and space. The network could provide a common storage area that could be referred to by collaborating pupils in different groups or from different classrooms. It could also allow different generations of pupils to leave tracks that, in turn, become a resource for new cohorts of learners and open up the possibility for the extension of education beyond the classroom. Through the use of computer technology, we can create new media for the organisation of common knowledge within communities of learners (Crook, 1996). Understanding what makes for successful collaborative interactions in these new environments will be challenging and will require an understanding of both the social and technical infrastructures that afford opportunities for productive interaction.

## 6 The Role of Social Comparison

All the approaches outlined thus far have been more or less concerned with the role of overt, observable processes of interpersonal interaction. However, there are studies that have yielded findings which challenge any notion that interaction and discussion alone underpin the beneficial effects of group work. For example, the work of Joiner and colleagues investigating children's computer-based problem solving has revealed that working in the presence of other children who are similarly engaged can be facilitative even if the children are working on different machines without any overt interaction (Joiner et al., 1995). We thus need to develop a fuller understanding of the more covert processes whereby the presence of other learners (and indeed adults) can influence a learner's response to the task in affective and motivational terms.

A relevant paper by Robinson-Stavely and Cooper (1990), studying computer-based learning in adults with very little prior computer experience, highlights the effects of the mere presence of a peer and emphasises the power of expectancy of success in influencing both performance and the response to others in a learning situation. In their first study, Robinson-Stavely and Cooper asked male and female students to complete a difficult computer task and a series of questionnaires in either the presence or the absence of another person. The male and female students were differentially affected by the presence of another person. Women who worked in the presence of someone else typically performed less well, reported higher anxiety and expressed more negative attitudes towards computers than women who worked alone. With the men, however, the presence of another person had the opposite effect. In a second study, Robinson-Stavely and Cooper manipulated the learners' expectations of success. Some students were given "feedback" that they were likely to do well and others that they were unlikely to succeed. There was evidence of facilitation for "positive expectancy" participants and impairment for "negative expectancy" participants, as compared with those participants who were required to work alone.

What emerges from the study of Robinson-Cooper and Staveley is the significance of social comparison rather than simply social facilitation in the learning situation. Social comparison (and, more generally, the role of social representations in learning) is central to research undertaken by Monteil (1991, 1992). Monteil's work has focused on social comparison variables affecting learning in secondary school pupils. In a series of studies generating clear and replicable results, Monteil and colleagues have manipulated actual academic ability levels, inducing expectancies of success in pupils relative to their peers on particular tasks, and varying the public versus private nature of the subsequent learning situation. Where there is a mismatch (whether positive or negative) between a student's general academic standing and their induced perception of their own ability relative to their peers, good learning outcomes are achieved in private or "anonymous" learning situations. However, those students who receive "feedback" matching their expectations learn more effectively in situations where their performance is visible to their peers (Monteil, 1992).

Monteil and colleagues interpret their results in terms of social comparisons within the peer-based learning situation, and interpret the effects of such comparisons as mediated by emotional responses to the situation, influencing the focusing of attention on task-relevant information. Their studies attest to the powerful influence of social representations (of self and of others) on children's learning. They also highlight the need to pay closer attention to the ways in which children working in the various conditions we create will construe the task, and to their own situation and ability in relation to it and their peers.

## 7 Concluding Remarks

As we have seen throughout this chapter, the advent of educational computer technology may not only promote new forms of collaborative activity amongst learners, but also illuminate the nature of our human capabilities as collaborative learners. We have also seen that studies based on the cognitive-developmental theories of Piaget and Vygotsky have taken us a considerable distance in understanding how and when peer interaction facilitates children's understanding. To get much further, we shall need to integrate a fuller understanding of the cultural context of peer interactions and learners' social perceptions and emotional responses with our cognitive accounts of "how group-work works". The challenge, then, is how best to develop an understanding of collaborative learning environments as systemic wholes where all the factors reciprocally affect each other (Salomon, 1994). Capturing complex processes in a schematised manner without simplifying them will represent a considerable challenge for researchers seeking to understand collaborative learning in computer environments.

# 3

# The Role of Grounding in Collaborative Learning Tasks

Michael Baker, Tia Hansen, Richard Joiner and David Traum

Collaborative learning tasks involve interaction between multiple participants, who thus need to maintain some degree of mutual understanding. The process by which this is accomplished is termed *grounding*. The way in which collaboration, grounding and learning take place is largely determined by the task, the situation and the tools available. This chapter discusses relations between grounding, collaboration and learning, drawing on research from two main areas: the language sciences and the cultural–historical activity theory (CHAT). We build a unifying perspective of mutual understanding mediated by material and semiotic tools that can be used for analysis as well as for design of collaborative learning tasks, especially those that are carried out via computer-mediated communication. We illustrate the perspective with reference to a particular computer-mediated collaborative learning situation in the domain of physics.

## 1 Introduction

Collaborative learning is a complex phenomenon that can be analysed at many different levels. One of the most basic dichotomies is between individual and social levels of analysis. From one perspective, all action and learning can be seen as individual—it is individuals who act and learn—and notions of group learning and action must be derived from these basic building blocks. Yet, collective action must be seen as more than just a sum of isolated individual acts. Moreover, as described in numerous studies (see, e.g., chapters 2 and 6, this volume), collaborative learning can be both quantitatively and qualitatively different from individual learning.

From another perspective, society or culture is more basic than the individual, having a continuing historical presence, gradually evolving as a consequence of the accumulated thinking and action of the individuals that it comprises. The main challenge here is to understand the specific nature of human learning that makes this accumulation of efforts possible. The answer suggested here is that human beings are capable of externalising (or "objectivising") thinking into tools (material tools, such as pens and computers, and semiotic tools, such as sign-systems). Such tools can then be appropriated by others (including new members of the culture, especially children), who can, in turn, further use and refine them. *Appropriation*—the special kind of learning that concerns such cultural objects—takes sign-mediated assistance from other members of the culture, who scaffold children's first attempts with the cultural object in such a way that they gradually move from being able to use tools under guidance to being able to use them on their own, and in their own way (Leontjev, 1981; core points also quoted in Tolman, 1988). In Vygotsky's words:

> "Every function in the child's cultural development appears twice: . . . first, *between* people (*interpsychological*), and then *inside* the child (*intrapsychological*)." (Vygotsky, 1978, p. 57)

Clearly, both of these perspectives are crucial to a comprehensive account of collaborative learning: to explain both how individual action can fruitfully combine to yield collaboration and learning, and how the cultural tools of a society can be appropriated and used by individuals through interaction within the society. The main aim of this chapter is therefore to explore the extent to which a theoretical perspective on collaborative learning can be constructed that unifies the two previously mentioned ones: "individual to society" and "society to individual". Given the ambitious nature of such an endeavour, we have chosen to focus on a single phenomenon that is central to the interplay between individuals and between individuals and society: the phenomenon of *grounding* (Clark & Wilkes-Gibbs, 1986).

*Grounding* is the name given to the interactive processes by which *common ground* (or *mutual understanding*) between individuals is constructed and maintained. Some mutual understanding between individuals will already exist at the start of any interaction, having been attained through integration of the individuals into a common culture. However, as an important part of the communication process, this common ground will need to be augmented with new information related to different facets of the activity, such as the individuals themselves, the tools, the goals and setting of the activity. Thus, during the interaction, as a result of grounding, learning may take place, by virtue of appropriation of refined tools. Our challenge is therefore to understand how these processes—grounding and appropriation—that operate on quite different timescales, lead to collaborative learning. In order to respond to this challenge, we need a deep understanding of the role that tools play in learning within cultures, together with micro-level analyses of how grounding actually takes place in the carrying out of concrete collaborative tasks. A unified perspective on the role of grounding in collaborative learning therefore needs to take into account the roles of both culture and interindividual interaction.

We draw on contributions from two main areas: *language sciences* research on grounding and related phenomena (e.g. Moeschler, 1985; Clark & Wilkes-Gibbs, 1986; Clark and Schaefer, 1989; Allwood et al., 1991; Roulet, 1992) and *cultural–historical activity theory* (CHAT) (e.g. Vygotsky, 1978, 1986; Leontjev, 1981; Cole & Engeström, 1993; Wertsch, 1994). Language sciences research provides fine-grained cognitive models of the grounding process, collaboration, and how the two relate, within the short timescale of isolated verbal interactions. CHAT, on the other hand, provides us with the notion of *learning as appropriation of tools*, and enables us to understand the role of language (the "tool of tools" according to Vygotsky) in collaborative learning. These two areas of research are thus complementary, given that grounding is a process of language interaction. Also, CHAT enables us to situate the role of verbal interaction, and grounding processes in particular, within a cultural–historical timescale.

Whilst each area of research has a separate contribution to make to our project, some pre-existing theoretical connections between the two are readily apparent. For example, the terms "intersubjectivity" (e.g. Forman, 1992), "negotiation of meaning" (e.g. Lave, 1991) and the "interpsychological plane of functioning" (e.g. Wertsch, 1994), referred to in CHAT-inspired areas of psychology, appear to refer to phenomena that closely relate to grounding, as described in language sciences.[1] Nevertheless, although these terms appear to designate similar phenomenon, we should not neglect the fact that the theories that underlie them are radically different: a deeper theoretical comparison is required if we are to benefit from both of these areas of research in understanding the role of grounding in collaborative learning.

In this chapter, we have concretised our theoretical perspective in the form of a simple framework (or model) for analysing collaborative learning tasks that highlights the different types of mutual understanding that need to be achieved for learning from collaboration, as mediated by tools. The framework is intended to be useful for designing collaborative learning tasks, particularly those that are computer-based or computer-mediated, since even the tools for interaction are, to some extent, under the designer's control. Throughout the chapter, we use a specific computer-mediated collaborative learning environment ("C-CHENE"—cf., Baker & Lund, 1997) in order to illustrate our theoretical perspective and to instantiate our analytical framework.

We begin by discussing the elements of our object of study (grounding, collaboration, learning) and the possible relations between them from the two different theoretical perspectives: language sciences and CHAT. Section 2 gives a brief overview of work in languages sciences on grounding. Section 3 then considers how learning takes place, on a more macro-level, involving appropriation of tools. Next, we consider how to combine these two areas of work to achieve a perspective on the role of grounding in collaborative learning (section 4). We then present the analytical framework, which draws on each of these areas (Section 5), and describe an instantiation of it with respect to the C-CHENE computer-supported collaborative learning environment (section 6). We conclude by synthesising and developing our main claims, and making suggestions for future research on the role of grounding in computer-supported collaborative learning.

## 2 Grounding

A *common ground* of mutual understanding, knowledge, beliefs, assumptions, presuppositions, and so on, has been claimed to be necessary for many aspects of communication and collaboration. *Grounding* is the process by which agents augment and maintain such a common ground. Although agents who interact will usually already

1. For other recent attempts to integrate discourse analyses and CHAT, see, e.g., Wells (1996).

possess some such common ground—perhaps by virtue of their common membership of a particular culture or social group, their physical copresence or even due to their previous interactions—this common ground will also need to be augmented and maintained during the interaction itself, in order to take into account new aspects of the common situation or task.

### *2.1 Basic Principles*

Clark and Marshall (1981) pointed out that mere accessibility or presentation of information is not sufficient for it to be added to the common ground. Depending on the situation, other assumptions may be required, such as community co-membership, rationality, locatability, attention, and rationality. Clark and Schaefer (1989) went beyond this, claiming that feedback of some sort was needed to actually ground material in conversation, and that this grounding process was collaborative, requiring effort by both partners to achieve common ground. They point out that it is not necessary to fully ground every aspect of the interaction, merely that the participants attain the following "*grounding criterion*":

> "The contributor and the partners mutually believe that the partners have understood what the contributor meant to a criterion sufficient for the current purpose." (Clark & Schaefer, 1989, p. 262)

What actually counts as a "sufficient" criterion of mutual understanding depends, of course, on the reasons for needing this information in the common ground, and can vary with the type of information and the collaborators' local and overall goals. As discussed later in the chapter, the criterion may need to be particularly stringent if the collaborators' activity is to lead to *learning*.

The grounding process involves, in addition to the mentioning of facts and proposals in the presence of another, processes of diagnosis (to monitor the state of the other collaborator) and feedback. When things are going smoothly, this feedback can be just simple acknowledgements (or even implicit, by performing an action that is contingent on understanding). However, feedback can also take the form of *repairs* when understanding seems to deviate from commonality. Collaborators do not always explicitly attempt to check that everything is mutually understood. Rather, they often make *assumptions* as to what the common ground is and will become, on the basis that this will be more economical. Clark and Wilkes-Gibbs (1986) introduced the principle of *least collaborative effort*, claiming that conversational participants try to minimise the work that both do from the initiation of each contribution to its mutual acceptance. Thus, a common form of feedback is a simple signal that the addressee is listening and comprehending rather than a display of precisely how the addressee is understanding the speaker. When it becomes manifest that an assumption of shared understanding has turned out to be unwarranted, there will usually be some attempt to repair and firmly establish the common ground—though only where this is deemed to be worth the effort involved.

Many aspects of interaction may be grounded, not just the meanings of utterances. Allwood, Nivre and Ahlsén (1991) describe four basic communicative functions, which correspond to "levels" at which problems for maintaining common ground may arise (cf. also Clark, 1994):

- *contact* (whether the interlocutor is willing and able to continue the interaction);
- *perception* (whether the interlocutor is willing and able to perceive the message);
- *understanding* (whether the interlocutor is willing and able to understand the message);
- *attitudinal reaction* (whether the interlocutor is willing and able to react and adequately respond to the message—specifically, whether she/he accepts or rejects it).

Clearly, there is an ordering between functions: one cannot genuinely establish an attitudinal reaction unless the message is understood, which requires perception and, in turn, contact. In fact, Clark and Schaefer (1987) show that explicitly grounding on one of these level reveals that grounding at a higher level is not achieved. The process of proposal, diagnosis and feedback (which are central to grounding) can also be viewed as a kind of *negotiation,* where the different levels represent different objects of negotiation (Moeschler, 1985; Roulet, 1992; Baker, 1994). In fact, nearly all aspects of management of the interaction (see Bunt, 1995) may be negotiable (e.g. turn-taking, opening and closing the dialogue, focusing, time management).

Whether researchers speak of grounding, negotiation or intersubjectivity (see section 1), there is nevertheless common agreement that various forms of linguistic and non-linguistic feedback constitute the basic mechanisms by which the common ground is achieved and maintained. However, the precise form that these mechanisms take also depends on the *media* of communication concerned (Clark & Brennan, 1991)—a point that we return to in the subsequent discussion of the role of tools in learning (section 4.1).

Different media (e.g. face-to-face, telephone, video-teleconference, terminal teleconference, and e-mail) bring different *resources* to, and impose *constraints* on, grounding. For example, to what extent does the medium support *copresence* (ability to see the same things), *cotemporality* (messages received at the same time as sent), *simultaneity* (whether both parties can send messages at the same time or whether they have to take turns) or *sequentiality* (whether the turns get out of sequence)? These constraints lead to different associated "costs", in terms of the effort required of interactional participants. For example, how difficult is it to change speaker, to repair misunderstanding, to tell what is being responded to or to formulate what one wants to say?

Since different media have different combinations of these constraints and costs, one would therefore expect the principle of least collaborative effort (*re* the grounding criterion) to predict different styles of grounding for use in different media. Designers of computer-mediated communication systems, used for collaborative learning tasks, need to pay particular attention to these factors (see also chapter 9, this volume).

### *2.2 An Example of Grounding in Computer-Mediated Communication*

The short interaction sequence reproduced in Table 1 gives a concrete illustration of some of the points made earlier. It is taken from a corpus of computer-mediated interactions produced by two students using the C-CHENE system (Baker & Lund, 1997).

Extended sequences from the corpus will be analysed in more detail later in the chapter (see Table 3). For our present purposes, all that is necessary for understanding the example is to know that the students are attempting to draw a diagram (called an "energy chain") together on a shared computer screen, to represent energy in a simple experimental situation (a battery linked to a bulb by two wires). The diagram consists of reservoirs and transformers of energy, with transfers of energy that link them together.

First, with respect to grounding, the question arises as to how S2 could understand what S1 meant by "the transfer" (line 2)? The answer is: because S1 and S2 assume that they have a common ground. This common ground consists of a common memory for the problem-solving actions that have already been performed, which is partly externalised on the computer screen and must be partly remembered (they assume common knowledge of the fact that S2 just drew an arrow from the transformer to the reservoir, then deleted it). However, the students must also assume

Table 1: Extract from computer-mediated interaction using C-CHENE[a]

| *N* | *T* (s) | S | Dialogue | Trace of successive interface states/ actions |
|---|---|---|---|---|
| 1 | 208 | S2 | | 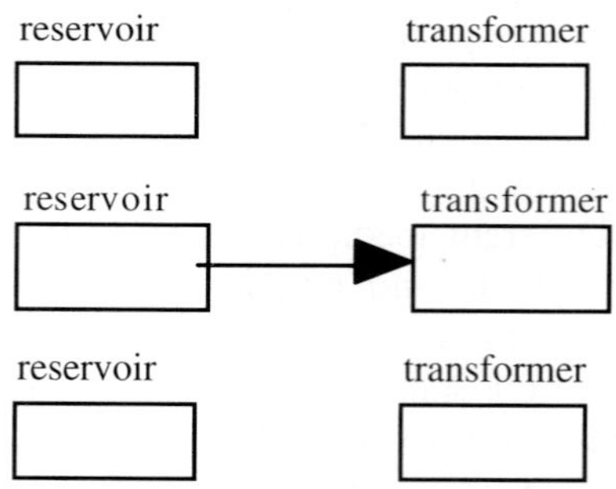 |
| 2 | 451 | S1 | I'll do the transfer | *(link from transformer to reservoir created then deleted)* |
| 3 | 482 | S2 | what about me | |
| 4 | 527 | S1 | you the second reservoir | |
| 5 | 564 | S2 | OK, go ahead | |
| | | | <...> | |

[a] *N* = line number; *T* (s) = time in seconds; *S* = speaker (S1) or (S2); Dialogue = typewritten synchronous messages (full screen-sharing). Only the dialogue history and the current state of the graphical interface are visible to both students.

common understanding that the arrow drawn and then deleted was intended to represent "a transfer". Furthermore, S2 needs these common assumptions in order to understand that S1 is talking about a transfer between the reservoir and the transformer. Just as S2 needs to access these assumptions in order to understand what S1 meant, S1 has in fact *formulated* and *presented the contribution* of line 2 on the basis of the same assumptions. Thus, S1 can produce his utterance with greater economy (*production costs* with this typewritten medium are high). Note that although these assumptions appear minimally sufficient for S1 to convey an understanding to S2 of what S1 is intending to do (draw an arrow from the reservoir to the transformer, to represent a transfer), and thus for the interaction to continue, this does not guarantee individual or mutual understanding in a "deeper" sense of what "transfer (of energy)" really is, in terms of the physics task (a point with which students often have difficulties).

Second, how does S1 know whether, or to what extent, S2 has understood the utterance of line 2, i.e. the extent to which the utterance is *grounded*? The case is quite complex, since no explicit feedback (e.g. "right", "OK, . . . ") is given. At this point in the interaction, the students are discussing "who will do what". Therefore, S2 demonstrates understanding of S1's utterance (line 2) in line 3 *by continuing relevantly* with respect to this common goal, i.e. S2 asks what she/he will do, as part of a negotiation of tasks (S2 will accept S1's claimed task, as long as S2 has an acceptable assignment).

A similar analysis can be given for lines 4 and 5. What is significant about line 5 is that S2 explicitly grounds the proposed division of responsibilities ("OK, . . .") on the level of *agreement*, thereby grounding *understanding* as well (agreement with $x$ presupposes understanding of $x$). At the end of this extract, the proposals for who will do what, with what, and when, can be assumed to be grounded to a sufficient degree for the participants in the interaction.

From this example, we can thus see: (1) some of the range of possible objects that can be grounded (referents such as "the transformer", meanings of words such as "transformer", responsibilities for and coordination of problem solving), (2) communicative levels at which grounding needs to take place (understanding, agreement) and (3) the role of the medium in the way that utterances are produced and grounded.

### 2.3 Computational Work on Grounding

Grounding is becoming increasingly important in computer dialogue systems. This is particularly true for systems with spoken language interfaces because current speech recognition technology produces far more errors than a human listener would make. In order for the system to interpret communications by the user correctly, the system must sometimes try to verify or attempt to repair its first hypothesis. As with human–human communications, there are a number of possible grounding strategies, ranging from explicit requests for verification, to demonstrating the current understanding (perhaps with a visual display), to implicitly revealing it by performing a contextually relevant action. Notions of the grounding criterion and principle of least collaborative effort apply for human–computer interaction as well: often, simple feedback from the user can be more efficient and accurate than computation. On the other hand, frequent requests for verification can be very annoying to the user, and will slow down the interaction. Much current work (e.g., Smith & Hipp, 1994; Danieli & Gerbino, 1995) is devoted to investigations of the best styles, i.e. what kind of verifications and when to perform them.

Traum (1994) presented a computational model of grounding, extending previous work by Clark and Schaefer (1989). In this model, utterances are seen as the performance of particular kinds of speech acts (such as initiate, continue, repair, and acknowledge) that change the state of groundedness of some information. The model allows one to form a precise theory (which can still turn out to be incorrect) of what is grounded and what actions need to be performed to achieve grounding at any point in the conversation. Moreover, the model included a relation of these acts to mental states of the agents in such a way that it could be implemented and used as a component in a human–computer dialogue system (Allen et al., 1994).

As well as providing the resource for a computational agent to engage in more flexible and natural grounding behaviour, such a model could also be used within a computer-mediated communication system to help make explicit to the users how their grounding of content is proceeding. Dillenbourg et al. (1997) describe preliminary implementations of "observer" agents that provide visual views of several measures of collaboration, including degree of grounding.

Computational work on grounding may thus be important in educational settings, for the design of computer-mediated communication in distance learning and for design of computer dialogue systems in computer-based learning systems.

### 2.4 *The Relations between Grounding, Collaboration and Collaborative Learning*

Having described grounding, we now discuss how it relates to collaboration and to collaborative learning. We argue that, under specific operational definitions of the terms "collaboration" and "collaborative learning", where the latter is interpreted as the learning that occurs by virtue of collaborative activity *per se*, grounding is an integral part of both.

As illustrated in the introductory chapter of this book (chapter 1, this volume), there are many notions of collaborative learning, and collaboration itself. Whilst various proposed definitions differ widely as to the specific necessary and sufficient ingredients of collaboration, most do converge on some relation to coordination and mutuality of the mental states of the collaborators. We will therefore use one recent and widely accepted definition of collaborative activity, proposed by Roschelle and Teasley:

> "[Collaboration is] a coordinated, synchronous activity that is the result of a continued attempt to construct and maintain a shared conception of a problem." (Roschelle & Teasley, 1995, p. 70)

This definition of collaboration, relying as it does on the attempt to construct and maintain a "shared conception", appears to be almost identical with the definitions of grounding and the common ground presented here. If this were so, then we could conclude that students are collaborating, quite simply, when they are engaged in grounding. However, we would claim that although grounding is a form of collaboration, and collaboration requires grounding, the two terms are not co-extensive. This is because grounding and collaboration can take place with respect to different objects and different communicative functions.

We claimed above (section 2.1) that grounding can take place with respect to four communicative functions—contact, perception, understanding and agreement—and with respect to different "objects"—meanings, propositions, rights, obligations, images, etc. When grounding and collaboration take place with respect to a specific level/object, then the two coincide. However, grounding and interaction may take place with respect to different levels/objects *simultaneously* in a given situation. In this case, there can be both collaboration and non-collaboration, but with respect to different levels/objects. Grounding and collaboration do not, therefore, necessarily coincide, although they may do so in certain cases.

Consider the case of two students collaborating, as in the example from Table 1. They will already possess a certain common ground as members of the same class, having worked together before, and so on. At one point, a disagreement over an intermediary solution to the problem could arise, be mutually recognised, and lead to an argumentative discussion. The students may be grounding and thus collaborating to the extent that each is trying to understand the other and to address the different points raised for or against the proposed solution. However, they may be competing rather than collaborating with respect to the communicative function of agreement on domain goals and propositions. More subtly, the domain-level disagreement may be accompanied by another disagreement, working on the level of the self-images that they attempt to impose in the interaction. For example, one may attempt to impose the image of being "more knowledgeable" or "more reasonable" than the other. In this example, there is thus coexistence of collaboration and its absence, but with respect to different levels/objects.

In summary, once we adopt the restricted notion of collaboration proposed by Roschelle and Teasley (1995), this being a definition that has become widely accepted, then students are collaborating when they are attempting to ground in some way, at some level. If we then adopt the premise that a useful concept of collaborative learning is the learning that occurs by virtue of the collaboration *per se*, then learning from grounding is "collaborative learning" in this sense.[2]

2. One may thus wonder whether the process of a student learning with the help of a more advanced tutor would be considered "collaborative learning." Clearly, the two can be pursuing the same goal—that the student learns—and will engage in grounding behaviour. While this might be termed a collaboration, we would not consider it to be collaborative learning.

## 3 Learning as Appropriation of Tools

Grounding, as described previously, is an important aspect of collaborating, and thus collaborative learning. Another crucial component to investigating the role of grounding in collaborative learning is, of course, learning itself. We now extend our theoretical perspective on collaborative learning to incorporate elements of CHAT, based on the work of Vygotsky (1978, 1986), Luria (1979) and Leontjev (1981). This enables us to take into account the role of tools and to extend the timescale of activity considered. Fundamentally, CHAT views learning and development as a socio-cultural activity where the child appropriates culturally accumulated knowledge with the assistance of other people and tools available in the setting.

### *3.1 Tool-Mediated Activity*

Within CHAT, the basic structure of human cognition is characterised as a triangle with the subject (S) and object (O) on two sides of its base and the medium or artefacts at its apex (M/A) (see Figure 1). In this model, the subject refers to an active, conscious agent, and the object represents the object of his/her cognition and action, which is the thing or phenomenon that he/she is thinking about and acting upon.

Natural or unmediated action is represented by the line between S and O, and refers to those capacities that the subject's own organism supplies, i.e. what the senses reveal, and the instincts that the subject has about what to do with an object of this kind. For example, a small child faced with a pot of boiling water senses an object with certain physical qualities (spatial qualities, colour, humidity, heat), follows his/her instinct to explore unknown objects and experiences new sensations of heat and pain. This is unmediated action. By unmediated action and cognition the child can learn that this object provides unpleasant sensations when touched. He/she may also be able to learn to distinguish between objects that have this quality and objects that do not. However, it takes mediated action to learn to understand why. The latter is a

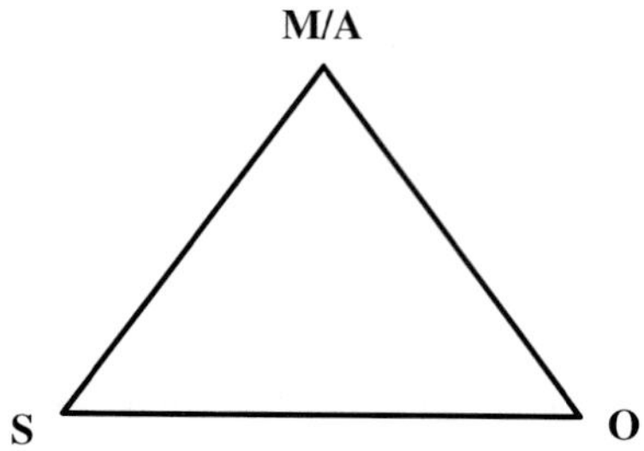


Figure 1: The mediational triangle (subject, object and medium/artefact)

theoretical issue, which presupposes ability to go beyond the immediacy of sensory experience.

Mediated action is depicted as the interaction of the subject with the object, mediated by auxiliary means (this is indicated by the lines S–M/A and M/A–O). Tool-mediated action refers not only to material tools (e.g. thermometers, computers, etc.), but also to semiotic ones (concepts, signs, and models). Language opens up a whole range of possibilities. Without language, people could deal only with those things that they could perceive and manipulate directly. With language, they can deal with things that cannot be perceived directly and with products of past generations' experience.

When a subject's actions are mediated, the task itself is fundamentally changed, relative to natural or unmediated action, and so is the cognition that goes with it. The task becomes a more complex structure of actions and cognitive processing since subjects must incorporate the three elements of the triangle and the possible relations between them into a unified idea in order to plan their actions and accomplish the task.

An example of action mediated by material tools would be when a child uses a thermometer to measure the temperature of a liquid. But even before she/he has learnt how to do that, she/he has already learnt to employ semiotic mediation in her/his quest for understanding. For example, she/he has begun to understand what is meant by a "boiling kettle". She/he has learnt words such as "boiling" and "temperature" and has a rudimentary understanding of what those concepts mean. Similarly, the example discussed in section 2 shows the use of two types of semiotic tools—the energy chain graphical "language' and, of course, typewritten language—that are embedded in specific material tools (computer interfaces).

Children are said to *appropriate* cultural objects (material and semiotic tools) when they learn from other members of the culture how those cultural objects are used, and what they are used to accomplish. Appropriation is not a process of rote-learning in which the individual simply adopts the facts and assumptions of the culture. Children appropriate these objects by participating in their use with more expert members of that culture. Learning results from the child's own experiences and practice with the object under the guidance of an expert. Thus, it is not a matter of information transmission from the expert to the novice, but of the novice "making this tool his own" (Leontjev, 1981).

### *3.2 Zone of Proximal Development*

Within the viewpoint of CHAT, the appropriation of semiotic tools must occur within the "zone of proximal development" (ZPD) (Vygotsky, 1978, 1986). This is defined as follows:

> "the difference between the actual developmental level as determined by independent problem solving and the level of potential development as determined through problem solving under adult guidance

or in collaboration with more capable peers." (Vygotsky, 1978, p. 86)

Children interacting in the zone of proximal development are participating in more-advanced tool-mediated activity than they are capable of doing independently, and in doing so they appropriate these cognitive tools, which advances their own independent problem solving.

Within the ZPD, the novice will not perceive the meaning of the object as fully as the expert user, and the novice and expert may have vastly different interpretations. An object such as a diagram or a representation on a computer interface may be understood very differently by the participants. Likewise, an utterance may be understood quite differently. This is not necessarily a problem (in reality, all utterances are always understood slightly differently by different interpreters; cf. Rommetveit, 1979, 1985). In some instances, it is not essential that the participants strive for sharedness of interpretations, as long as they agree enough to carry on the interaction (e.g. if they can point to the same place on the screen, they do not need the same interpretation of what is there for their attention to stay focused on that part of the interface). This ability to talk about objects that do not have a completely shared meaning is not a problem, but, in fact, an advantage. It allows for learning to occur when people with different perspectives interact. In order to untangle differences in understanding, participants are forced to make their assumptions explicit, to argue, to reason about and to exemplify them. In order to reach agreement, they need to construct a conception that allows for apparently insightful assumptions from both parties to be integrated, which will often require a conceptualisation that is more complex than the individuals" original ones.

### *3.3 Intersubjectivity*

Thus, an important process in the social origins of learning is joint construction of mutual understanding, or intersubjectivity (Rommetveit, 1979, 1985; Forman 1992). Wertsch (1985) identifies four levels of intersubjectivity in the transition from interpsychological functioning to intrapsychological functioning in an interaction between a mother and a child. The task was to copy a puzzle (a truck) in accordance with a model. Wertsch characterised the core goal-directed action in the adult's situation-definition as a series of three steps:

1. Consult the model to determine the identity and location of the next piece.
2. Select the piece identified in step 1 from the pile of pieces.
3. Add the piece selected in step 2 to the copy.

The first level of intersubjectivity is characterised by the fact that the child's situation-definition is so different from the adult's and that this makes combination very difficult. The adult may try to direct the child, but the child's understanding of the setting is so limited that the child does not interpret the adult's utterances appropriately. For example, one child shifted his attention to the real windows in the room in response to the mother's comment to look at the truck window.

At the second level, the child has a limited understanding of the setting. The child shares the adult's basic understanding of the objects in the setting (i.e. they depict pieces of a truck). However, the child does not fully understand the relationship that has been constructed between the truck and the model, and will often fail to respond appropriately to adult directions. For example, in response to one mother asking the child "to put the wheel on next", the child picked up the wheel and placed it in the wrong place, not taking in account of where the wheel was on the model.

At the third level, the child can respond to the adult's directions. She/he is making the appropriate inferences to the adult's statements, which indicates that interpsychological functioning is beginning to account for much of the child's performance. The adult does not have to specify all the steps, because the child has a fairly complete situation-definition. At the fourth level, the child takes over complete responsibility for carrying out the task.

Wertsch (1985) characterises learning as the transition from inter- to intrapsychological functioning. This process involves negotiating an intersubjective situation-definition with the tutor. Initially, this will be performed on the basis of a primitive intersubjectivity with the tutor and, subsequently, by going through a number of situation redefinitions until finally the child acquires a mature culturally appropriate situation-definition that provides the basis for self-regulation. Children working with adults can interact at a more advanced level of intersubjective understanding than they would be capable of independently.

In summary, from the vantage point of CHAT, learning can be seen as the appropriation of cultural tools. Grounding (on some level, with respect to some object) is thus necessary, but clearly not sufficient for collaborative learning (as we define it). First, if students are to learn from grounding, the common ground, or "background" of material and semiotic tools provided by culture, must be already appropriated by students to some degree. This is precisely what is usually left out by approaches that focus exclusively on the internal dynamics of interpersonal interactions. Second, collaboration provides a vehicle for learning if the individual participates in solving a task that is beyond his or her present autonomous ability. However, there are constraints on how far collaboration can promote cognitive abilities at any given point: grounding, or "intersubjectivity", must be situated within the ZPD if it is to lead to learning.

## 4 Towards Synthesis: Grounding and Collaborative Learning as the Appropriation of Tools

In the previous two sections, we presented some basic material necessary for understanding grounding as an interactive process and the CHAT theory of learning as the appropriation of tools. We now discuss some points of convergence and divergence between these two areas of research, with a view to understanding how grounding could lead to collaborative learning.

### *4.1 Language as a Tool and an Object of Grounding in Communication*

The most obvious convergence between the grounding model and the notion of learning as appropriation of tools obtains via the notions of *language* and *communication*. Grounding is a process whose primary function is to ensure effective communication in the sense of shared understanding of utterances in interaction. Although non-verbal signals (e.g. gestures) and situational factors (such as copresence) play an important role in this process, the primary objects that are themselves to be grounded are, in fact, linguistic productions (utterances). The extent to which interactions are linguistically or else non-verbally based is a function of how much common ground is already established: when participants do not know each other very well, they will need to speak; when they have interacted together for 50 years, a slight raising of an eyebrow may be sufficient for communication. So, we can conclude that grounding in the collaborative learning situations that we consider here is primarily language-based.

From the point of view of the activity theorist, language is a (semiotic) tool, a mediator of activity and thinking. During their development, children learn to use language (in speech) as a tool in different ways: to enlist help from others by communicating with them, to shape their ongoing activity, to plan their future activity, to structure their reasoning, and so on (Vygotsky, 1978). They may also refine the meanings of words in order to use them as more elaborated tools in new situations—e.g. they may learn to elaborate the meaning of the word "heat" from its use in everyday life to its use in the science classroom. Language is thus a multiform tool: it enables interpersonal communication and it is instrumental in individual and collective activity and thinking.

How does this theory of language as a tool transform our way of seeing learners who are engaged in grounded interaction ? Three main points of convergence can be identified.

First, the instrumental view of language as a tool (for guiding action, exchanging information, planning, and so on) leads us to concentrate on the interactional context, the *discourse genre*, in which grounding takes place, and to ask the following question: *why* are the learners grounding and what are their main *goals* ? This gives us a way of understanding the grounding criterion mentioned earlier (section 2.1), where it is stated that participants attempt to attain mutual understanding to a degree that is deemed sufficient *for the current purpose*. An understanding of learners' purposes (goals, motivations) is therefore crucial to understanding their grounding activity. In the next section, we discuss how these goals may not only be local to the problem-solving task, but also be largely determined by the subjects' interpretations of the *situation* in which they find themselves.

Second, language—a tool—may itself be an *object* of mutual understanding. For example, learners may be led to attempt to understand a language itself (e.g. foreign language learning) or to inquire about the "deeper" meaning of a particular term in their common language. This view thus leads us to ask the question, in addition to the above: *what* are the students grounding? The answer from activity theory in this case (language) is semiotic tools. These include symbolic, iconic and natural languages.

The type of learning especially associated with grounding will therefore be appropriation of the very medium with which grounding takes place. We discuss such a formal language, whose appropriation constitutes learning in a specific domain of physics, in section 6 of this chapter.

Third, communication in language may sometimes involve the use of material (communication) tools. In spoken face-to-face interactions, this is less evident, but is more so in the case of computer-mediated communication. In terms of the grounding model, such tools are considered as the communication "channel". However, such a metaphor could lead one to think of the channel as a passive medium, however much "noise" is introduced. There is thus an important convergence between the notion of material tools, constituting the physical communication medium, as *mediators* that shape communication and the results concerning the role of media in grounding mentioned earlier (section 2.1). Thus, communication "costs" and "constraints" within the grounding model can be understood as difficulties in appropriating material tools, thereby influencing the way in which semiotic tools are used and appropriated in grounding.

Viewing learners engaged in grounding as appropriating tools leads us to focus on understanding learners" goals within specific types of discourse, on what is being grounded and on how the material tools used for communication itself shape the grounding process.

### 4.2 Grounding, Intersubjectivity and Learning as Appropriation of Tools

What is the type of learning associated with grounding—at least from a theoretical point of view—and what are the necessary conditions for its occurrence? We mentioned above that CHAT leads us to pose the important question of learners' goals that underlie their grounding activity, and the focus of it. To deal with the second question first—what learners are attempting to ground—from the point of view of the grounding model (and see, e.g. Schwartz, 1995), we make the following conjecture: *collaborative learning will be associated with a relatively high grounding criterion, with a high degree of cognitive-interactional effort.*

However, for learning to take place in a particular knowledge domain, this effort needs to be directed towards the right object. In simple terms, we can say that grounding can take place on pragmatic and semantic levels.[3] At the pragmatic level, grounding involves, for example, interactional participants understanding each others' communicative intentions—what the other is "trying to tell them". At the semantic level, it involves searching for common understanding of referents and meanings of terms. Pragmatic-level grounding is part of *learning to collaborate*: one learns to understand generally what the other will be trying to tell us. Semantic-level grounding, however, relates to attaining mutual understanding of what is meant by

3. Note that the semantics/pragmatics distinction is a subject of much controversy in language sciences—see, e.g. Allwood (1976) and Levinson (1983) for reviews of these debates.

certain terms and expressions; it thus relates more closely to learning in a specific knowledge domain by means of interpersonal interaction. Learning to understand one another and learning to understand semiotic tools are thus two important aspects of collaborative learning as we define it. It seems plausible that students will need to have learned how to collaborate in order to learn together in a task domain. We can thus view learning by grounding as a gradual transition from pragmatic to semantic grounding.

Finally, from our above discussion, it should be clear that what has been termed semantic grounding can also be viewed as appropriation of a semiotic tool. Our previously mentioned conjecture can thus be reformulated as follows: *collaborative learning will be associated with a gradual transition from the use of language as a medium for grounding communication (pragmatic) to grounding on the level of the medium itself (semantic), leading to appropriation of the medium.*

Why should students expend extra effort in semantic grounding, rather than contenting themselves with a level of pragmatic grounding that enables the interaction to continue in a minimal sense? In this theoretical discussion, we can point only to the necessity for detailed analysis of learners' goals, their motivations, as mentioned above. Some tasks may be intrinsically motivating for some students; with other students, extrinsic situational goals may come to the forefront. It also appears plausible that learners must have some understanding of what they are trying to do, or are "supposed" to do, in order to be willing to expend effort on trying to understand the task: what researcher or teacher has not had the experience of discovering subsequently that the subjects/students had no clear understanding of what they were doing or had been asked to do? A more concrete question is: what are the conditions that would favour such grounding activity? Some answers may be found in the research on intersubjectivity, referred to earlier (section 3.3). On one hand, semantic (or "deep") grounding cannot lead to learning if the concepts dealt with are simply beyond the students'' zone of proximal development. On the other hand, it seems that an optimal level of difference between learners' conceptual viewpoints can stimulate the desire to gain deeper understanding.

### *4.3 Recapitulation*

Grounding is a process that is directed primarily towards attaining mutual understanding of linguistic productions (utterances). Language is thus the primary medium of grounding and (in virtue of its role as a semiotic tool) may itself be the object of attempted mutual understanding. Learning from grounding—collaborative learning—can thus be viewed as appropriation of semiotic tools, mediated by those very tools. Interactants' goals in specific situations determine the extent to which they will be willing to expend effort to achieve mutual understanding. The learning associated with grounding will be associated with a high degree of expended cognitive-interactional effort to gain mutual understanding. Appropriation of cognitive tools can be viewed as a gradual transition from pragmatic-level grounding—learning to understand each other and to collaborate—towards learning to understand the semiotic tools in a specific domain (that may also be languages themselves). Apart from intrinsic and extrinsic (situational) motivations, learning from grounding can only occur within the zone of proximal development and may be stimulated by an optimal level of differences between individual perspectives. Gaining deep conceptual understanding takes an amount of time that extends well beyond interactions themselves and which draws on interactants'' understandings of tools inherited from culture.

We conclude that *understanding the role of grounding in collaborative learning requires a unit of cognitive analysis that includes agents, tools and goals in situation, together with relations of understanding between them.* In the next section, we present an analytical framework that encapsulates this synthesis.

## 5 A Framework for Analysing the Role of Grounding in Collaborative Learning Tasks

We now present a simple framework for analysing the role of grounding in collaborative learning tasks. The framework highlights the different forms of mutual understanding that will need to be achieved and gives a central place to tools as both mediators of understanding and objects of it. It may thus be used as an aid to designers of collaborative learning tasks, and particularly those that are performed via computer-mediated communication. In this case, designers have to actually to decide how to engineer communicational and interactional possibilities for potential learners, and so must take into account the costs and opportunities that these channels provide for achieving different types of mutual understanding.

According to our framework, a collaborative learning task is defined by the pursuit of certain goals by a group of agents via the use of specific tools in a specific situation. This goal-oriented activity requires the achievement of different types of understanding and mutuality.

The basic elements of the analytical framework are shown in Figure 2. The nodes in the diagram in Figure 2 represent the principal entities involved—two agents (or sets of them), tools, the agents' goals and the situation in which these goals are collaboratively pursued. Lines represent a single type of relation between entities—tool-mediated individual or mutual understanding:[4]

- $\alpha_1/\alpha_2$–$\gamma$, $\Sigma$: agents' (tool-mediated) understanding of the goals/situation;
- $\alpha_1$–$\gamma$, $\Sigma$–$\alpha_2$: agents' (tool-mediated) mutual understanding of the goals/situation;
- $\alpha_1$–$\alpha_2$: agents' (tool-mediated) mutual understanding of each other (each other's actions/utterances);
- $\alpha_1/\alpha_2$–$\tau$: agents' (tool-mediated) understanding of the tools themselves;
- $\alpha_1$–$\tau$–$\alpha_2$: agents' (tool-mediated) mutual understanding of the tools themselves.

4. For convenience, the relation is represented here simply by "–". Elsewhere (see, e.g. Table 2), we represent directionality of understanding as follows: "$x \rightarrow y$" ($x$'s understanding of $y$) and "$x \leftrightarrow y$" ($x$'s and $y$'s mutual understanding of each other). When it is important to make explicit the role of rools as mediators of the understanding relation, we represent this as follows (see, e.g. section 6 of this chapter): "$x$–[$\tau$]–$y$".

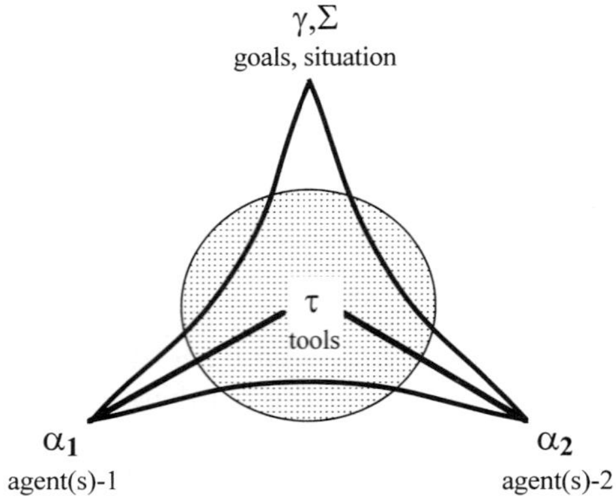


Figure 2: A framework for analysing the role of grounding in collaborative learning tasks

Thus, tools ($\tau$) are a special case since they have a dual role: they are both *objects* of understanding (straight lines from $\alpha_1$ and $\alpha_2$ to $\tau$) since learners have to understand the tools themselves (material tools, such as communicative media for producing messages and manipulating the environment, and semiotic tools, such as language), and *mediators* of agents' understanding of each other, the task and the goals-in-situation (the curved lines that pass through the circle around $\tau$). The agents' goals ($\gamma$) and the situation ($\Sigma$) in which they are pursued have been represented as a unique yet complex entity since the two may be viewed as inextricably linked (see the discussion below).

The symbols for the entities in the diagram of Figure 2 should be viewed as variables whose specific instantiations can give rise to representations of a large variety of collaborative learning tasks. *Learning* is not represented explicitly in the diagram, since neither is *time*. *Learning may be achieved when the entities and relations change throughout time.*

### 5.1 *Agents* ($\alpha_1, \alpha_2$)

These are the distinct entities that collaborate. At a very general level, they could be quite heterogeneous kinds of entities, such as individual people, groups, or even corporate entities. Some computer programs can also be viewed as agents (although this view is not without its critics[5]). The basic criterion is that each agent can meaningfully be conceived as having its own mental state and ability to act and interact with the other entities represented in the model. Each agent should be able to communicate with the other, and represent the mental state of the other (at least enough to engage in a grounding process). In addition, agents should be able to conceptualise

5. See, e.g. Agre (1995) for a review of research on artificial agents in artificial intelligence (AI), and Winograd and Flores (1986) for a well-known critique of this AI view.

and act within the situation towards achievement of the goals and be able to use tools. Agents should also be able to learn about these aspects and be held responsible (to the other agent) for specific actions, including communicative ones.

Even with respect to prototypical agents—single human beings—there is a large variety of types of agents, particularly when one considers their relationship to each other. Agents could be adults or children, could have equal or unequal abilities, could be novices or accomplished at the tasks and using the tools, could be familiar or unfamiliar with each other, could be in an equal or hierarchical social relation with each other, and so on.

### 5.2 *Goals and Situation* ($\gamma, \Sigma$)

What social-interactional situation are the agents embedded in, and what are they trying to do? What are the constraints on what they can do? Several distinctions need to be made.

The first is between *external* and *internal* (or "subjective") goals. Clearly, the goals of a collaborative learning task, as determined by a teacher or an experimenter, are not necessarily the same as those perceived and pursued by the learner agents. Within agents' internal goals, the possibility of conflict between them is an important factor to be taken into account.

Second, agents' goals can be considered as relating to the concrete activity to be performed (e.g. acquiring understanding of a concept or a fact, solving a problem) or as situational (e.g. the goal of pleasing the teacher, or getting a good mark in an educational institution). Such situational goals are not necessarily "inferior" to those that directly relate to the task to be achieved. In fact, the situational goals may be exploited in school settings as vehicles for initiating children in activities that, when adequately supported, may lead to concentration on the task goals themselves (Leontjev, 1981).

Social-institutional situations may be viewed as imposing a set of constraints, as may concrete activities or problems. Both need to be taken into account, but for the purposes of this chapter it is not useful to make a sharp division between the two; thus, we have treated goals and situations as a single, yet composite entity in the framework.

Third, any specific instantiation of the model presupposes an understanding of what is to count as a "situation". Thus, from one point of view, a problem solved during a psychology experiment is not a "situation" to the extent that the social relationship between the experimenter and subject, and the latter's understanding of the external goals (see above), are not taken into account. From another point of view (see, e.g. Perret-Clermont, Perret & Bell, 1991), an experimental situation is a social situation *per se*, given the fact that the experimenter will necessarily represent some social role or position from the point of view of the subject.

In general, situations will be multifaceted, and there can be several goals for each agent or group of them, along the distinctions outlined above.

### *5.3 Tools ($\tau$)*

As mentioned above, tools are both *objects* of (mutual) understanding and *mediators* of it. There are few *a priori* limits to what can actually be a tool: what makes something a tool or not is the role that it plays in activity itself. On the other hand, there is a crucial distinction in types of tools in that sometimes an object becomes a tool in a particular situation, in an *ad hoc* manner, whereas other objects are *created* to be tools for *particular kinds of activities*. It is, of course, the latter type of tools that are complex cultural achievements, and which accordingly require instruction in order to be appropriated.

We consider two main types of tools: *material* tools (such as hammers and computers) and *semiotic* tools (such as natural or formal languages). The role of each is well illustrated by the case of computer-mediated collaborative work. Here, the technology is on one hand, of course, a material tool: it allows concrete tasks to be achieved, including the task of communicating. On the other hand, the technology is also the support for semiotic tools such as natural language and other formal languages that may be incorporated in the task environment (see section 6 of this chapter).

Finally, it is to be noted that, within our model, tools (or any other entity) are not to be considered in isolation from other entities: they are not mere "add-ons" to humans; the unit of analysis is person(s)-acting-with-mediational-means, viewed as an indissociable whole (Mammen, 1993).

### *5.4 Relations in the Framework: Processes of Understanding and Learning*

A single type of relation between entities in the model is taken into account: (tool-mediated) *understanding*. There are, in fact, two forms of it: '"agent $\alpha$'s understanding of $x$", and "agents $\alpha_1$'s and $\alpha_2$'s mutual understanding of $x$". In both cases, "$x$" can also be another agent or a group of them. We do not give here a formal definition of *understanding,* relying for our present purposes on common-sense intuitions. Of course, "understanding" will be different for different types of agents. While each of the following sentences is meaningful in an appropriate context, the word "understanding" does not capture precisely the same relationship between agent and understood concept: "John understands the energy chain model", "The computer understands what John meant in drawing a box", "Volkswagen understands how to sell cars".

The relations of understanding between agents $\alpha_1/\alpha_2$ and the goals-in-situation pass *exclusively* via mediating tools. This means that, in our model, we exclude consideration of understanding that is *not* mediated by tools.

We also exclude the case of a single solitary agent understanding by perception and action (even though this is also arguably mediated by semiotic tools), since our model is restricted to situations where several agents collaborate. In this case, it is clear that nearly all understanding will be mediated by tools, especially when communication is involved. Given our own definition of collaborative learning described earlier, all understanding is mediated by semiotic tools, and sometimes by material tools. In fact, the grounding process, as modelled by Clark and Schaefer (1989) and Traum (1994), whose constituents are speech acts such as acknowledgements, verifications, and repairs, is itself a prime semiotic tool for achieving mutual understanding.

As pointed out earlier, *learning* may take place when the entities and relations of the model undergo *change* through *time*. Usually, we will be concerned with changes that are viewed as positive from some normative point of view (what is usually understood as learning: Johnson-Laird, 1988, p. 129). Theoretically, therefore, the following types of change and learning are possible in collaborative learning tasks (see Table 2).

In most collaborative learning situations, it is assumed that the tools, goals and situations do not undergo change, often because the *timescale* considered is usually little more than an hour. In our framework, we assume a *complex dynamic system*, in which all entities and relations between them can change. In the case of material tools or social situations, for example, this will usually involve considering a much longer timescale (e.g. for a computer system or for educational practices to change), in accordance with CHAT (see Cole & Engeström, 1993). Thus, the necessity to include consideration of a wider timescale radically changes the types of understanding that need to be taken into account in the collaborative learning situation under study. Once we consider a broader timescale, we need to consider changes in goals of institutions and of the tools that are elaborated and appropriated. On a micro-level, grounding is a prime mechanism by which these global changes take place.

Table 2: Types of change (and learning) in the framework for analysing collaborative learning tasks[a]

| Change (model syntax) | | Explanation/example |
|---|---|---|
| Entities | $\Delta\alpha_i$ | Cognitive change in an agent (learning, unlearning, acquiring "correct" or "incorrect" knowledge or beliefs) |
| | $\Delta\tau$ | Change in material or semiotic tools (e.g. introduction of a new computer system, language development) |
| | $\Delta\gamma, \Sigma$ | Change in agents' goals-in-situation (e.g. pedagogical goals may change in educational institutions) |
| Relations | $\Delta\alpha_1 \leftrightarrow \alpha_2$ | Change in mutual understanding between agents |
| | $\Delta\alpha_1 \rightarrow \alpha_2/\alpha_2 \rightarrow \alpha_1$ | Change in an agent's understanding of another agent |
| | $\Delta\alpha_i \rightarrow \tau$ | Change in agents' understanding of tools |
| | $\Delta\alpha_i \rightarrow \gamma, \Sigma$ | Changes in agents' understanding of goals and situations |

[a] "$\rightarrow$", "$\leftrightarrow$" represent relations of tool-mediated understanding and mutual understanding respectively; $\Delta x$ represents change in $x$.

Changes in one entity or a relation can thus effect changes in another, and this may be bi-directional. The specific causal relation with which we are principally concerned here is, of course, collaborative learning, a cognitive change in an agent produced as a result of a change in the agent's mutual understanding during interaction: $\Delta\alpha_1 \leftrightarrow \alpha_2 = causes => \Delta\alpha_i$. The causal arrow may, however, be reversed since changes in agents effected by other means may trigger the necessity to attain new mutual understanding. There may be several cycles of this as agents oscillate between understanding and non-understanding (Miyake, 1986). Changes in mutual understanding may also concern the tools, goals and situations. Similarly, when tasks and goals change over longer periods of time (either as represented by agents or as a result of changes in a social situation, such as within a school), then this may require and cause changes in agents' understanding of each other.

Mapping out the relevant entities and understanding involved in collaborative learning tasks, and studying the possible range of interdependencies between changes, thus enables us to consider a wider range of types of learning than is usually taken into account. In addition, it provides a way of modelling collaborative learning situations within which grounding and collaboration processes are integrated within use of material and semiotic tools.

## 6 An Illustrative Analysis: A Computer-Supported Collaborative Learning Environment for Modelling Energy in Physics

We now illustrate the use of our framework in analysing the use of a specific computer-mediated collaborative learning (CSCL) environment—C-CHENE[6] (Baker & Lund, 1997). The environment was designed for learning about modelling in physics, and the theory/model of energy in particular (Tiberghien, 1994, 1996; Devi et al., 1996). We have chosen to analyse this environment since the task it supports provides a good illustration of the complexity of material and semiotic tools and because it has been experimented with using two different computer-mediated communication interfaces. The latter allows us to compare the different grounding mechanisms provided by the interfaces and to suggest how they could be improved in further research.

### *6.1 The Energy Modelling Task*

In the part of the energy modelling teaching sequence discussed here, students are asked to produce qualitative models (diagrams) called "energy chains" in order to represent energy storage, transfer and transformation in simple experimental situations (the task will be discussed in more detail throughout the ensuing analysis). Figure 3 shows an example energy chain (correct) for an experimental situation

6. "CHENE" = "CHaîne ENErgétique" = "energy chain"; "C-CHENE" = Collaborative CHENE.

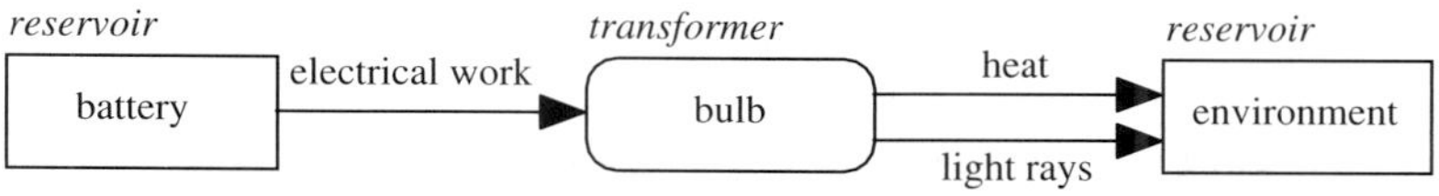


Figure 3: Energy chain (correct) for an experiment where a bulb is connected to a battery by two wires

where a battery is linked to a bulb by two wires (see later Table 3 for an example of a student energy chain for this experiment).

The didactic rationale of this task is as follows: by attempting to instantiate the simple qualitative model for energy in concrete cases, the students will be led to co-construct an understanding of the concepts of "energy" and "model".

In the studies referred to here, the students worked in pairs, and at a distance via a network, each on their own separate computers but sharing and working on the same virtual space. Energy chains were drawn using a customised graphical interface and all (typewritten) communication during the task execution passed via specially designed communication interfaces.

### 6.2 The Agents ($\alpha_1$ and $\alpha_2$)

The students who participated in the studies were 16–17 years old, and all came from the same class of a secondary school (lycée) in Lyon. Their teacher asked for volunteers who had some facility with typing on a computer keyboard. The students grouped themselves into "friendship pairs".

### 6.3 The Goals and the Situation ($\gamma, \Sigma$)

In any learning situation, we clearly cannot assume that the goals that designers of the situation intend learners to have will genuinely be perceived and adopted by them. We therefore distinguish the following: the overall research goals of the task designer, the designers' goals with respect to what the students are intended to learn, and the goals that the students appeared to be following (from the researchers' analysis).

The overall research goals of the studies were to understand the cognitive processes of modelling used by students and to understand how students collaborate in performing such tasks, focusing on the role of computer-mediated communication. The designer's learning goals for students were as follows. As a result of creating energy chains for the three experiments, students should gain a better understanding of (1) the concept of energy in physics (e.g. energy can be neither created nor destroyed; it does not correspond to a "tangible" entity like a physical object, such as a battery, nor to one of its electrical properties, such as resistance); and (2) the nature of modelling in science (e.g. there is no necessary one-to-one matching between model and reality, models select certain phenomena but not others, they have an internal syntax, and so on).

The task was not invented solely as a research tool, but also with the educational goals that it should be used in actual classrooms for learning about energy (it is currently being incorporated in the French physics teaching National Curriculum). The educational goal of the study was therefore also to test the feasibility of a part of a complex teaching sequence in a computer-mediated communication setting. In technological terms, we also had the goal of testing the usability of a relatively new version of the C-CHENE software.

From analysis of students' interactions with C-CHENE (Baker & Lund, 1997), the students' goals appear to have been the following: (1) to manage using a piece of complex software, with which they were previously unfamiliar; (2) to draw an energy chain that satisfied the energy chain model rules; (3) to draw an energy chain that both of the pair agreed upon; and (4) to have a friendly discussion.

Students were not always concerned with getting the "right answer" (although, rarely, some did obtain it), nor with understanding energy modelling. Rather, they often aimed to produce a solution that was minimally satisfactory to each, and to get through the task with minimum effort, especially with minimum (typewritten) communication. The most important researcher goal—that the students should learn what energy and modelling are—was rarely explicitly recognised (i.e. discussed or adopted). The fact that the task was carried out outside school hours, in a research institution without teacher presence or assessment, is undoubtedly important in explaining the students' activity.

### *6.4 The Tools (τ)*

The principal tools provided and used for the task are both material (computer software and hardware) and semiotic. Into the latter category, as well as natural language (typewritten), the energy chain graphical language is of particular interest, as is the fact that the semiotic tools are embedded in specific material tools.

#### 6.4.1 Material Tools (Technology)

Collaborative problem-solving interactions with two different communication interfaces were compared (see Figure 4 below): (1) a "chat-box" interface and (2) a "dedicated" interface.[7]

In the first case, students type their messages in their own boxes, then "send" them (tab key), which causes the sender's box to close and that of the other student to open. The message sent is added to the end of a continually visible interaction history. The upper part of the screen is where energy chains are drawn, with specially adapted

7. Only the communication buttons are shown for the dedicated interface in Figure 4, redrawn and translated from the original French. The energy chain construction interface remains the same as with the chat-box interface.

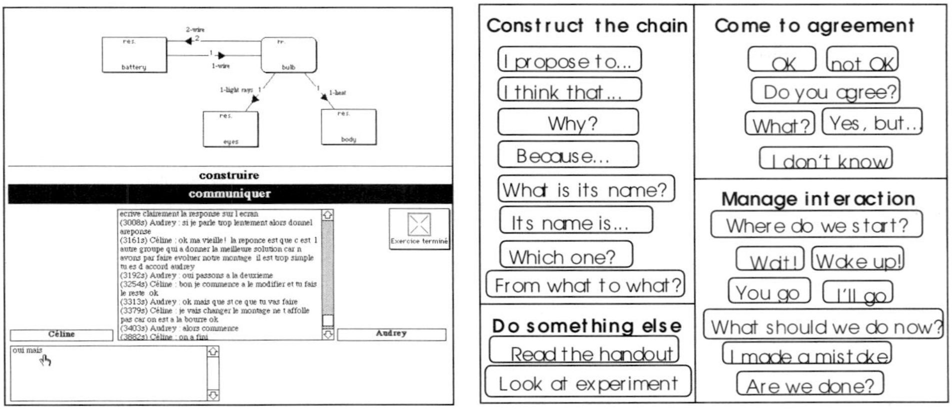


*"Chat-box" interface* *"Dedicated" interface*

Figure 4: Two communication interfaces of the C-CHENE CSCL environment: left, "chat-box"; right, "dedicated"

menus; change between construction and communication modes is done with the two central buttons that go across the screen. The idea of such a customised design was to visually reinforce the coordination constraints here: communication and construction cannot be carried out simultaneously, and only one student can communicate at a time.

The dedicated interface provides buttons for specific speech acts required for performing this task (see Baker & Lund, 1997, for design considerations). Again, the interaction history is visible. The principle behind this interface was that the speech act buttons would alleviate the typing burden, freeing up effort for the energy chain task itself, and that the presence of buttons such as "because" and "why?" would encourage students to reflect on their solutions, during discussion, instead of simply drawing their solution. Specific buttons are also provided for facilitating *grounding*: "OK", "not OK", "I don't know", "do you agree?".

### 6.4.2 Semiotic Tools (Typewritten Natural Language and the Energy Chain Language)

In addition to the natural language used by the students to communicate, the energy chain model is also a semiotic tool. It was not a pre-existing cultural object since it was created as a didactic simplification, whose use by students in the solving of specific modelling problems would lead them to co-construct understanding of the concepts of energy and of modelling itself.

The "lexicon" of the model is a set of boxes, arrows and labels with determinate meanings within the model itself. It has a simple set of syntactic constraints (e.g. "start and end with a reservoir", "no more than three arrows between two boxes"). The didactic aim is then that students will co-construct the *semantics* of the language—the intermediary concepts necessary for understanding the nature of the relations between model and reality (Tiberghien, 1994).

The specific way in which this semiotic tool is embedded in the material tools may have important effects on students'' cognitive processes. For example, when drawing energy chains on paper, students could decide that "the bulb gives out light", and so draw an arrow out from the bulb without specifying where it goes to. This is not possible on the graphical interface: the arrow cannot go in mid air, it must be linked to another box. So, in this case, students must draw a box (unlabelled at present), deciding already whether it is a transformer or a reservoir. They are therefore obliged to make a more global matching of a sub-part of an energy chain. Although such global matching represents a more complex modelling process, it is not clear whether imposing it from the start will facilitate learning. The important point to be recognised is that such constraints need to be considered explicitly in designing educational tools.

### *6.5 Relations in the Framework: Grounding and Learning*

We discuss relations in our analytical framework—what needs to be grounded and how it is grounded—with respect to two example interactions, produced by pairs of students with the chat-box and dedicated interfaces, respectively, of C-CHENE.

#### 6.5.1 Example Interaction Sequences

Table 3 shows a complete interaction of 30 minutes duration (see also section 2.2) produced by two female students for the battery–wires–bulb experiment, using the chat-box interface of C-CHENE.

The first 11 minutes of the interaction shown in Table 3 (lines 1–11; note, the initial part of the dialogue is not shown), during which a reservoir for the battery and a transformer for the bulb were created, can be compared with the following extract from the beginning of a dedicated-interface interaction for the same task (Table 4), of approximately 10 minutes duration. Only the part up to where the students create a reservoir for the battery and a transformer for the bulb is shown, for sake of brevity.

#### 6.5.2 Tool-Mediated Grounding of the Students' Interaction ($\alpha_1$–[$\tau$]–$\alpha_2$)

The two interactions shown in section 6.5.1 differ markedly in the form of the students' collaboration (Baker, 1995)—who will act (graphically), who will communicate, when and with respect to what—and consequently, in the way that grounding takes place.

In the case of the chat-box interaction (Table 3), the students' interaction appears to be *graphical-action-based*, communication being reduced to a minimum (e.g. the interaction even begins by S1 drawing two rectangles without previous discussion), this being one way of reducing coordination problems with the interface. Instead of attempting to co-construct the solution, on the basis of a more reflective discussion, the students adopt collaboration pattern that reduces to "you draw, I will tell you what I think". Within this restricted form of collaboration, coordination is also negotiated explicitly from the outset of the interaction (lines 2–5). *Grounding* therefore often takes place via a combination of perception of graphical action and communication, but which is not always successful in maintaining mutual *understanding* (e.g. line 16: "what are you doing"). The students do, however, seem concerned to attain mutual *agreement:* they ask whether the other agrees (line 12) and when they have performed a graphical action without previously negotiating who will act, they explain their actions (line 18).

The form of collaboration and the way in which grounding takes place with the dedicated interface interaction (Table 4) is, by contrast, much more *discussion-based.* The students begin by discussing a common strategy, by proposing a solution element that will not be drawn before common agreement is reached (see, e.g., lines 3–10 in Table 4). The form of collaboration is "discuss possible solution, reach agreement, draw solution". As well as explicit grounding acts on the *agreement* level (lines 8, 9, 13, 15, 18, 20), the students also ground the interaction along the levels of *contact* (line 5) and *understanding* (lines 16, 17).[8]

With both interfaces, however, although grounding can take place via coperception of graphical actions, students also had to perform some activities individually—they each had a paper copy of the energy model and their own experimental apparatus—in which case, there was no common perception. This obliged the students to make explicit what they were doing, and what aspect of the text they were talking about, thus adding to communication costs. In future designs of the system, these activities therefore need to be coperceivable. On the other hand, the presence of a commonly perceivable interaction history may have helped in focusing the interaction, this being one of the possible advantages of computer-mediated communication over face-to-face audio communication.

### 6.5.3 Students' Understanding of the Tools ($\alpha_1$–$\tau$)

The forms of collaboration and grounding that were produced with each interface appear to be linked to the extent to which students are able to *appropriate (learn) the material tools*—the computer-mediated communication interfaces. Given that their communication was typewritten, producing linguistic feedback in grounding involves a higher production cost with the chat-box interface than with the dedicated one (clicking on a button). This may, in turn, explain why a more discussion-based form of collaboration was adopted, accordingly, with the dedicated interface.

The students were given a 30-minute training session prior to the study, in which they were supposed to learn to use the physical tools (computer interfaces). Clearly, this was insufficient, since students experienced coordination problems during the study. In fact, it is difficult to see why students should invest much effort in learning

8. See discussion of Allwood, Nivre and Ahlsén (1991), in section 2.1 of this chapter.

Table 3: Interaction with chat-box interface of C-CHENE, students S1 and S2, battery–wires–bulb experiment[a]

| *N* | *T* (s) | S | Dialogue | Interface state/actions |
|---|---|---|---|---|
| 1 | 208 | S2 | |   |
| | | | | *(link transformer -> reservoir created then deleted)* |
| 2 | 451 | S1 | I'll do the transfer | |
| 3 | 482 | S2 | what about me | |
| 4 | 527 | S1 | you the second reservoir | |
| 5 | 564 | S2 | OK, go ahead | |
| 6 | 591 | S1 | |   |
| 7 | 617 | S1 | over to you | |
| 8 | 707 | S2 | but you have to start with a reservoir and finish with a reservoir | |
| 9 | 803 | S1 | I began with a reservoir then a transformer now I have to put a reservoir OK | |
| 10 | 828 | S2 | OK | |
| 11 | 868 | S1 | | 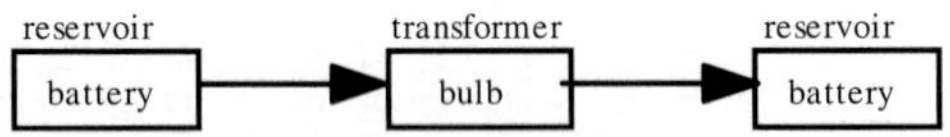  |

| | | | | |
|---|---|---|---|---|
| 12 | 1173 | S1 | do you think it's right, what I did with our diagram | |
| 13 | 1259 | S2 | it's right but there has to be a generator or something like that | |
| 14 | 1309 | S1 | the battery is the generator! banana brain! | |
| 15 | 1369 | S2 | | 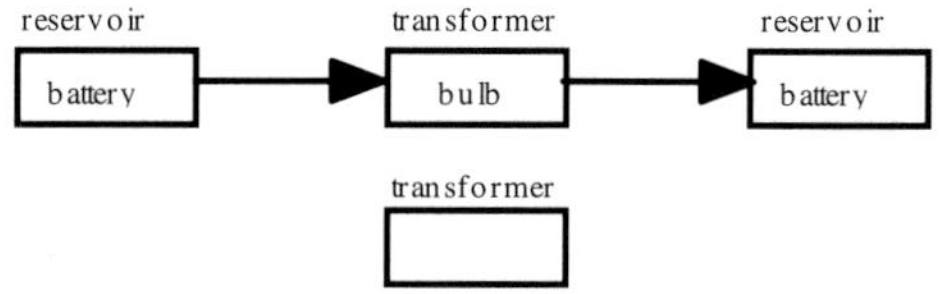  |
| 16 | 1408 | S1 | what are you doing | |
| 17 | 1530 | S1 | | 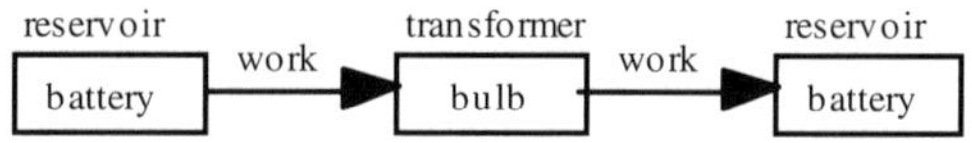  |
| 18 | 1715 | S1 | I've named the transfers because that's what's asked for on the sheet about transfers! look! | |
| 19 | 1760 | S2 | you and your errors so give me a break all the time's going, we have to finish | |
| 20 | 1830 | S1 | It's finished now OK let's call the old lady OK my dear! | |
| | | | *** end of interaction *** | |

[a]The interaction has been translated from the original French. $N$ = line number; $T$ (s) = time from start of session in seconds; S = speaker (S1 or S2)

Table 4: Dedicated-interface interaction (extract) for battery-bulb experiment[a]

| *N* | *T* (s) | S | Dialogue | Interface state/actions |
|---|---|---|---|---|
| 1 | 56 | S1 | <Read the sheet> | |
| 2 | 95 | S1 | <Where do we begin?> | |
| 3 | 121 | S2 | <I propose to . . . > <create a reservoir> | |
| 4 | 144 | S1 | <Which one?> | |
| 5 | 179 | S1 | <Hello there!> | |
| 6 | 200 | S2 | <I don't know> | |
| 7 | 213 | S1 | <I think that . . . > the first reservoir is called battery | |
| 8 | 262 | S1 | <Do you agree?> | |
| 9 | 287 | S2 | <Agreed> | |
| 10 | 312 | S1 | | reservoir<br>[battery] |
| 11 | 357 | S1 | <What shall we do?> | |
| 12 | 376 | S2 | <( propose to . . . > <create a reservoir> | |
| 13 | 397 | S1 | <Not agreed> | |
| 14 | 412 | S1 | <I propose to . . . > <create a transformer> | |
| 15 | 432 | S1 | <Do you agree?> | |
| 16 | 457 | S2 | <Why?> <14> | |
| 17 | 472 | S1 | <I think that . . . > the bulb is a transformer of electrical energy | |
| 18 | 450 | S2 | <Agreed> | |
| 19 | 554 | S1 | <I think that . . . > I'm going to create the transformer called bulb | |
| 20 | 604 | S2 | <Agreed> | |
| 21 | 645 | S1 | | reservoir [battery] transformer [bulb] |
| | | S2 | < . . . > | |

[a]Text between "< . . . >" corresponds to an automatic tracing of the speech act buttons that are clicked. All other text is typed by the students themselves. Numbers such as "<14>" refer to line numbers of the automatic transcription, as referred to by the students (e.g. line 16 above). Again, the example is translated from the original French.

the tools, since they knew that the tools were research prototypes that they would be unlikely to encounter subsequently.

The specific semiotic tool used here—the energy chain model—presented few learning difficulties, at least at the surface level. At a deeper, semantic level, learning about this tool constituted one of the major goals of the task, from the researchers' point of view, i.e. the desired learning here may be viewed as *appropriation of a semiotic tool.* Interestingly, students did not necessarily view this tool as "fixed", even though it was provided from an authoritative source (the researcher): they sometimes elaborated its meaning.[9]

### 6.5.4 Students' Understanding of Goals-in-Situation ($\alpha_1$–$\gamma$, $\Sigma$–$\alpha_2$)

As mentioned above, the students' main goals with the chat-box interface interaction rarely appeared to match with those of the researchers: the latter being to acquire a deeper understanding of energy and modelling. The students' main goals appeared to be *learn how to use the tools*, *draw a diagram* (solve the explicitly given task) and *maintain social interaction*. With the dedicated interface interaction, however, students did engage in explanation of the nature of energy (see, e.g. lines 16, 17 of Table 4). Later on in that specific interaction, more extended explanation sequences were produced that more closely approached the researchers' goals.

There are two main lines of explanation for these facts. The first is that the cognitive load associated with learning the material tools was sufficiently high so as to exclude the kind of reflection that would be associated with gaining deeper understanding of the domain. Thus, the students tried to minimise coordination problems associated with switching between communication and construction. The second relates to the novelty of the situation in which the students found themselves. Given that the task took place outside a school environment, during free time and without teacher assessment, this meant that little was actually at stake for the students. They therefore expended little effort with the task beyond learning the material tools themselves and minimally drawing a diagram.

All these factors contribute to explaining the students' apparent goals as a function of learning to understand each other and the tools. The interest of such an *a posteriori* analysis is therefore that it highlights how the situation, the goals and the tools require redesigning so as to bring researcher and student goals closer together.

## *6.6 Reflections: How Does Analysis with the Model Inform Redesign?*

Above all, the analysis reveals the importance of *tools*—material and semiotic—in mediating understanding and learning within a computer-mediated communication situation.

9. An example is described in Baker (1996): one student implicitly elaborated the meaning of the energy chain rule "A complete energy chain must start and end with a reservoir" to include " . . . and the first and last reservoirs can not correspond to the same object or event in the experimental situation".

First, the physics task can be viewed as one of appropriation of a semiotic tool—the energy chain model. In future versions of C-CHENE, we therefore need to reconsider carefully the way in which this cognitive tool is embodied in material tools so that the constraints that are imposed are those that are most likely to foster modelling processes in students' problem solving and that are at an appropriate level of complexity for them at a given point in their problem solving.

Second, the communication (material) tools provided have a marked effect on grounding. Although some progress was made in facilitating grounding in language interaction with the second dedicated interface, other aspects of the task (energy model rules, experimental apparatus) need to be made mutually perceivable during communication to reduce grounding costs.

Third, the difficulties involved in learning to use the communication tools may be important in explaining the mismatch between researcher and student goals. Clearly, if we want students to learn energy and modelling concepts, we have to reduce the cognitive-communicational load required in learning the tools, and in learning about the task.

Finally, we need to directly address what the students' understanding of the task situation is, so that, even outside school constraints, there is something at stake for them in problem solving.

## 7 Concluding Remarks

Our main aim in this chapter was to explore synergies between two areas—language sciences research on grounding, and CHAT research on learning—as a means of furthering our theoretical understanding of collaborative learning. This exploration gave rise to a simple framework for analysing and designing collaborative learning situations (especially those that involve computer-mediated communication) that highlights the different forms of grounding that need to be taken into account.

From the point of view of understanding the cognitive effects of communicative interactions produced in teaching–learning situations, the CHAT view of learning as appropriation of tools appears particularly appropriate, given its emphasis on the role of language, the "tool of tools", in interpersonal interactions and in thinking. From the point of view of CHAT, micro-analyses of the interactive processes at work in grounding, as well as the different planes on which it can operate, may be of potential use in understanding precisely how the interpersonal plane of functioning within the ZPD leads to appropriation of tools.

The overall picture of collaborative learning that emerges from this research may be summarised as follows: collaborative learning is associated with the increased cognitive-interactional effort involved in the transition from learning to understand each other to learning to understand the meanings of the semiotic tools that constitute the mediators of interpersonal interaction. Understanding how and when grounding leads to collaborative learning therefore requires detailed analysis of learners' *goals* in specific *situations* that motivate them to go further in their attempt to gain mutual understanding, of the extent to which the *material and semiotic tools* that they use can be appropriated, and, especially, of the different forms of mutual understanding that will need to be achieved by the learners with respect to these "objects". It is precisely these aspects of collaborative learning situations that our analytical framework is designed to highlight.

We have proposed a first step in a possible longer term research project that aims to elaborate a new unifying theoretical framework for understanding collaborative learning. Although a much more far-reaching theoretical integration remains to be achieved, we nevertheless hope that this work will provide a basis for further research from the point of view of each of the contributing disciplines. As with any such framework, an important test of its usefulness will be the extent to which it can give rise to the design of effective collaborative learning situations in the future, particularly those that involve computer-mediated communication.

## Acknowledgements

This research was carried out within the framework of the European Science Foundation (ESF) research programme "Learning in Humans and Machines", to which we are grateful for financial support. David Traum was partially supported, while writing this paper, by the Swiss National Science Foundation under Research contract 1114–040711.94, the U.S. Army Research Office under contract/grant number DAAH 04 95 10628 and the U.S. National Science Foundation under grant IRI-9311988. Tia Hansen was supported by the Danish Council for Research in the Humanities under grant no. 9502778. We would like to thank members of the GRIC-COAST research team who participated in the studies analysed in this chapter and Bob Lewis, Pierre Dillenbourg, Karen Littleton, Ulrich Hoppe, Daniel Schwartz and Yrö Engeström for comments on earlier versions of this work.

# 4

# What is "Multi" in Multi-agent Learning?

Gerhard Weiss and Pierre Dillenbourg

The importance of learning in multi-agent environments as a research and application area is widely acknowledged in artificial intelligence (AI). Although there is a rapidly growing body of literature on multi-agent learning, almost nothing is known about the intrinsic nature of, and requirements for, this kind of learning. This observation is the starting point of this chapter, which aims to provide a more general characterization of multi-agent learning. This is done in an interdisciplinary way from two different perspectives: the perspective of single-agent learning (the "machine learning perspective") and the perspective of human–human collaborative learning (the "psychological perspective"). The former leads to a "positive" characterization: three types of learning mechanisms—multiplication, division, and interaction—are identified and illustrated. These can occur in multi-agent but not in single-agent settings. The latter leads to a "negative" characterization: several cognitive processes, such as mutual regulation and explanation, are identified and discussed. These are essential to human–human collaborative learning, but have been largely ignored so far in the available multi-agent learning approaches. Misunderstanding among humans is identified as a major source of these processes, and its important role in the context of multi-agent systems is stressed. This chapter also offers a brief guide to agents and multi-agent systems as studied in artificial intelligence, and suggests directions for future research on multi-agent learning.

## 1 Introduction: What is Intrinsic to Multi-agent Learning?

Learning in multi-agent environments[1] had been widely neglected in artificial intelligence until a few years ago. On the one hand, work in distributed artificial intelligence (DAI) mainly concentrated on developing multi-agent systems whose activity repertoires are more or less fixed; and on the other hand, work in machine learning (ML) mainly concentrated on learning techniques and methods for single-agent settings. This situation has changed considerably and, today, learning in multi-agent environments constitutes a research and application area whose importance is broadly acknowledged in AI. This acknowledgment is largely based on the insight that multi-agent systems typically are very complex and hard to specify in their dynamics and behavior, and that they therefore should be equipped with the ability to self-

1. Throughout this chapter, the phrase "multi-agent" refers to artificial agents as studied in artificial intelligence (as in e.g. "multi-agent environment" or "multi-agent learning") and does not refer to human agents. Human agents will be explicitly mentioned when comparing work on multi-agents with psychological research on collaborative learning.

improve their future performance. There is a rapidly growing body of work on particular algorithms and techniques for multi-agent learning. The available literature, however, has almost nothing to say about the intrinsic nature of, and the unique requirements for, this kind of learning. This observation forms the starting point of this chapter, which is intended to focus on the characteristics of multi-agent learning, and to take the first steps toward answering the question "What is 'multi' in multi-agent learning?"

Our approach to this question is interdisciplinary, and leads to a characterization of multi-agent learning from two different perspectives:

- single-agent learning ("*machine learning perspective*") and
- human–human collaborative learning ("*psychological perspective*").

The ML perspective is challenging and worth exploring because, until some years ago, learning in multi-agent and in single-agent environments were studied independently of each other (exceptions simply prove the rule). As a consequence, only very little is known about their relationship. For instance, it is still unclear how exactly a computational system made of multiple agents learns differently compared with a system made of a single agent, and to what extent learning mechanisms in multi-agent systems differ from algorithms as they have been traditionally studied and applied in ML in the context of single-agent systems. The psychological perspective is challenging and worthy of exploration because research on multi-agent learning and research on human–human collaborative learning grew almost independently of each other. Despite the intuitively obvious correspondence between these two fields of research, it is unclear whether existing multi-agent learning schemes can be observed in human environments, and, conversely, whether the mechanisms observed in human–human collaborative learning can be applied in artificial multi-agent systems.

The structure of this chapter is as follows. Section 2 provides a compact guide to the concepts of agents and multi-agent systems as studied in (D)AI. The purpose of this section is to establish a common basis for all subsequent considerations. This section may be skipped by the reader who is already familiar with the (D)AI point of view of these two concepts. Section 3 then characterizes multi-agent learning by focusing on characteristic differences in comparison with single-agent learning. Section 4 characterizes multi-agent learning by contrasting it with human–human collaborative learning. Finally, section 5 concludes this chapter by briefly summarizing the main results and by suggesting potential and promising directions of future research on multi-agent learning.

## 2 A Brief Guide to Agents and Multi-agent Systems

Before focusing on multi-agent learning, it is useful and necessary to say a few words about the context in which this kind of learning is studied in DAI. This section offers a brief guide to this context, and tries to give an idea of the AI point of view of agents and multi-agent systems. The term "agent" has turned out to be central to major developments in AI and computer science, and today this term is used by many people working in different areas. It is therefore not surprising that there is considerable discussion on how to define this term precisely and in a computationally useful way. Generally, an agent is assumed to be composed of four *basic components*: usually called the sensor component, the motor component, the information base, and the reasoning engine. The sensor and motor components enable an agent to interact with its environment (e.g. by carrying out an action or by exchanging data with other agents). The information base contains the information an agent has about its environment (e.g. about environmental regularities and other agents" activities). The reasoning engine allows an agent to perform processes like inferring, planning and learning (e.g. by deducing new information, generating behavioral sequences, and increasing the efficiency of environmental interaction). Whereas this basic conception is commonly accepted, there is ongoing controversy on the characteristic properties that let an object like a program code or a robot be an agent. Following the considerations in Wooldridge and Jennings (1995b), a weak and a strong notion of agency can be distinguished. According to the *weak notion*, an agent enjoys the following properties:

- autonomy,
- reactivity and
- pro-activeness.

According to the more specific and *strong notion*, additional properties or mental attitudes such as

- belief, knowledge, etc. (describing information states),
- intention, commitment, etc. (describing deliberative states), and
- desire, goal, etc. (describing motivational states),

are used to characterize an agent. Examples of "typical" agents considered in (D)AI are shown below. The reader interested in more details on AI research and application on agency is particularly referred to Wooldridge and Jennings (1995a), Wooldridge, Müller and Tambe (1996), and Müller, Wooldrige and Jennings (1997).

In artificial intelligence, recent years have shown a rapidly growing interest in systems composed of several interacting agents instead of just a single agent. There are three major reasons for this interest in multi-agent systems:

1. They are applicable in many domains that cannot be handled by centralized systems.
2. They reflect the insight gained in the past decade, in disciplines like artificial intelligence, psychology and sociology, that intelligence and interaction are deeply and inevitably coupled to each other.
3. Recently, a solid platform of computer and network technology for realizing complex multi-agent systems became available.

Many different multi-agent systems have been described in the literature. According to the basic conception and broadly accepted viewpoint, a multi-agent system consists of several interacting agents that are limited and differ in their motor, sensory and cognitive abilities, as well as in their knowledge about their environment. We briefly illustrate some typical multi-agent scenarios.

One widespread and often used scenario is the transportation domain (e.g. Fischer et al., 1993). Here, several transportation companies, each having several trucks, transport goods between cities. Depending on the level of modeling, an individual truck, coalitions of trucks from the same or different companies, or a whole company, may correspond to and be modeled as an agent. The task that has to be solved by the agents is to complete customer jobs under certain time and cost constraints. Another example of a typical scenario is the manufacturing domain (e.g. Peceny, Weiss & Brauer, 1996). Here, several machines have to manufacture products from raw material under certain predefined constraints, like cost and time minimization. An individual machine, as well as a group of machines (e.g. those machines responsible for the same production step), may be modeled as an agent. A further example is the loading dock domain (e.g. Müller & Pischel, 1994). Here, several forklifts load and unload trucks according to some specific task requirements. Analogous to the other domains, the individual forklifts, as well as groups of forklifts, may correspond to the agents. As a final example, the toy-world scenario known as the predator/prey domain should be mentioned (e.g. Benda, Jaganathan & Dodhiawala, 1986). Here, the environment consists of a simple two-dimensional grid world, where each position is empty or occupied by a predator or prey. The predators and prey (which can move in a single-step mode from position to position) correspond to the agents, and the task to be solved by the predators is to catch the prey—for instance, by occupying their adjacent positions. Common to these four examples is that the agents may differ in their abilities and their information about the domain, and that aspects like interaction, communication, cooperation, collaboration, and negotiation are of particular relevance to the solution of the task. These examples also give an impression of the three *key aspects* in which multi-agent systems studied in DAI differ from each other (Weiss, 1996):

- the environment occupied by the multi-agent system (e.g. with respect to diversity and uncertainty),
- the agent–agent and agent–environment interactions (e.g. with respect to frequency and variability) and
- the agents themselves.

The reader interested in available work on multi-agent systems is referred to the Proceedings of the First and Second International Conference on Multi-Agent Systems (1995, 1996, see references). Examples of standard literature on DAI in general are Huhns (1987), Bond and Gasser, (1988), Gasser and Huhns (1989) and O'Hare and Jennings (1996).

## 3 The ML Perspective: Multi-agent versus Single-agent Learning

As mentioned above, multi-agent learning is established as a relatively young but rapidly developing area of research and application in DAI (e.g. Imam, 1996; Sen, 1996; Weiss & Sen, 1996; Weiss, 1997). Whereas this area had been neglected until some years ago, today it is commonly agreed by the DAI and the ML communities that multi-agent learning deserves particular attention because of the inherent complexity of multi-agent systems (or of distributed intelligent systems in general). This section characterizes and illustrates three classes of mechanisms that make multi-agent learning different from single-agent learning: *multiplication, division* and *interaction*. This classification provides a "*positive*" *characterization* of multi-agent learning (we later present also a "negative" classification). For each mechanism, a general "definition" is provided and several representative multi-agent learning techniques are briefly described.

This coarse classification does not reflect the variety of approaches to multi-agent learning. Each class is an abstraction of a variety of algorithms. Most work on multi-agent learning combines these mechanisms, and therefore a unique assignment is not always possible. This also means that the three classes are not disjointed, but are of different complexity and partially subsume one another: divided learning includes elements of multiplied learning, and interactive learning includes elements of divided learning. Our goal was not to compare all existing work in DAI, but to isolate what is very specific to multi-agent learning and has not been investigated in single-agent learning. There are, of course, a few exceptions, since there is, for instance, something "multi" in research on parallel inductive learning (e.g. Chan & Stolfo, 1993; Provost & Aronis, 1996) or on multistrategy learning (e.g. Michalski & Tecuci, 1995).

## *3.1 Multiplication Mechanisms*

### 3.1.1 General

If each agent in a multi-agent system is given a learning algorithm, the whole system will learn. In the case of multiplied learning, there are several learners, but each of them learns independently of the others; i.e. their interactions do not impact on their individual learning processes. There may be interactions among the agents, but these interactions just provide input that may be used in the other agents' learning processes. The learning processes of agents but not the agents themselves are, so to speak, isolated from each other. The individual learners may use the same or a different learning algorithm. In the case of multiplied learning, each individual learner typically pursues its own learning goal without explicitly taking notice of the other agents' learning goals, and without being guided by the wish or intention to support the others in achieving their goals. (The learning goals may mutually supplement each other, but this is more an "emerging side-effect" than the essence of multiplied learning.) In the case of multiplied learning, an agent learns "as if it were alone", and in this sense has to act as a "*generalist*", capable of carrying out *all* activities that as a whole constitute a learning process.

At the group level, the learning effects due to multiplication may be related to variables such as the number of agents or their heterogeneity. For instance, if the system includes several identical robots, it is less dramatic if one of them breaks down (technical robustness). If the system includes agents with different background knowledge, it may work in a wider range of situations (applicability). The effects of multiplication are salient in sub-symbolic systems or systems being active below the knowledge level; this includes a large number of elementary learning units (which, however, can hardly be called agents). Good examples of such systems are artificial neural networks composed of large numbers of neurons.

### 3.1.2 Examples of "Multiplied Learning"

Ohko, Hiraki and Anzai (1996) investigated the use of case-based learning as a method for reducing communication costs in multi-agent systems. As an application domain, a simulated environment occupied by several mobile robots is used. The robots have to transport objects between predefined locations, where they differ from each other in their transportation abilities. Each robot has its own case base, where a single case consists of the description of a transportation task and a solution, together with its quality. Several agents may learn (i.e. retrieve cases and detect similarities among cases) concurrently, without influencing each other in their learning processes. Although the robots interact by communicating with each other in order to announce transportation jobs and to respond to job announcements, the learning processes of different agents do not interfere with each other.

Haynes, Lau and Sen (1996) used the predator/prey domain in order to study conflict resolution on the basis of case-based learning. Here, two types of agents—predators and prey—are distinguished. According to this approach, each predator can be thought of as handling its own case base, where a single case consists of the description of a particular predator/prey configuration and a recommended action to be taken in this configuration. It is clear that predators interact by occupying the same environment and by sensing each others' position. Despite this, however, a predator conducts its learning steps independently of other predators and their learning steps (even if the different learning processes may result in an increased overall coherence).

Vidal and Durfee (1995) developed an algorithm that allows an agent to predict other agents" actions on the basis of recursive models: i.e. models of situations that are built on models of other situations in a recursive and nested way. The models of situations are developed on the basis of past experience and in such a way that costs for development are limited. Here, learning is multiplied in the sense that several agents may concurrently learn about each other according to the same algorithm without requiring interaction, except mutual observation. There is no joint learning goal. The learning processes of multiple learners do not influence each other, and they occur independently of each other. Observation just serves as a "mechanism" that provides the information upon which a learning process is based.

Terabe et al. (1997) dealt with the question of how organizational knowledge can influence the efficiency of task allocation in multi-agent settings; i.e. to assign tasks to agents in such a way that the time for completing the tasks is reduced. Organizational knowledge is information about other agents' ability to solve tasks of different types. In the scenario described by Terabe and his colleagues, several agents can learn independently of each other how specific tasks are best distributed across the overall system. There is no explicit information exchange among the agents, and the only way in which they interact is by observing each other's abilities. The observations made by an agent are used to improve its estimate of the abilities of the observed agent, but the improvement itself is done by the observer. Each agent responsible for distributing a task conducts the same learning scheme, and none of the responsible agents is influenced in its learning by the other agents.

Carmel and Markovitch (1996) investigated the learning of interaction strategies from a formal, game-theoretic perspective. Interaction is considered as a repeated game, and another agent's interaction strategy is represented as a finite state-action automaton. Learning aims at modeling others' "interaction automata" based on past observations such that predictions of future interactions become possible. As in the research described above, the learning of the individual agents is not influenced by the other agents. The agents interact by observing each other's behavior, and the observations then serve as an input for separate learning processes. There is also no shared learning goal; instead, each agent pursues its own goal, namely, to maximize its own "profit". Related, game-theoretic work on multi-agent learning was presented by Sandholm and Crites (1995) and Mor, Goldman and Rosenschein (1996).

Bazzan (1997) employed the evolutionary principles of mutation and selection as mechanisms for improving coordination strategies in multi-agent environments. The problem considered was a simulated traffic flow scenario, where agents are located at the intersections of streets. A strategy corresponds to a signal plan, and evolution occurs by random mutation and fitness-oriented selection of the strategies. There are several other papers on learning in multi-agent systems that follow the principle of biological evolution (for instance, see Bull & Forgarty, 1996; Grefenstette & Daley, 1996; Haynes & Sen, 1996). These are essentially examples of the multiplied learning approach because the agents do not influence each other in their learning processes. It has to be stressed that a classification of evolution-based learning is difficult insofar as "evolution", reduced to the application of operators like mutation and selection, can be also viewed as a centralized process. From this point of view, the agents just act, whereas "learning by evolution" is something like a meta-activity that takes place without being influenced by the agents themselves.

## 3.2 Division Mechanisms

### 3.2.1 General

In the case of "divided learning", a single-agent learning algorithm or a single learning task is divided among several agents. The division may be according to functional aspects of the algorithm and/or according to characteristics of the data to be processed in order to achieve the desired learning effects. As an example of a functional division, one could consider the task of learning to optimize a manufacturing process: here, different agents may concentrate on different manufacturing steps. As an example of a data-driven division, one could think of the task of learning to interpret geographically distributed sensor data: here, different agents may be responsible for different regions. The agents involved in divided learning have a shared overall learning goal. The division of the learning algorithm or task is typically carried out by the system designer, and is not a part of the learning process itself. Interaction is required for putting together the results achieved by the different agents, but, as in the case of multiplied learning, this interaction does only concern the input and output of the agents' learning activities. Moreover, the interaction does not emerge in the course of learning but is determined *a priori* and in detail by the designer. This is to say that there are interactions among the participants, but without a remarkable degree of freedom (i.e. it is known *a priori* what information has to be exchanged, when this exchange has to take place, and how the exchanged information has to be used). An individual agent involved in "divided learning" acts as "*specialist*", just responsible for a specific sub-set of the activities that as a whole form the overall learning process.

Different benefits can be expected from a division of labour. The fact that each agent only computes a sub-set of the learning algorithm makes it simpler to design and reduces the computational load of each agent. The "agent" metaphor insists on the autonomy of functions with respect to each other. A benefit expected from this increased modularity is that the different functions may be combined in many different ways. A further potential benefit is that a speed-up in learning may be achieved (provided that the time gained by parallelism is not outweighed by the time required for coordination). The main difference between division and multiplication is *redundancy*: in division, each agent performs a different sub-task while in multiplication, each agent performs the same task. In other words, the rate of redundancy in agent processing would be 0% in "pure" division mechanisms and 100% in "pure" multiplication, with the actual design of existing systems being, of course, somewhere in between these two extremes.

### 3.2.2 Examples of "Divided Learning"

Sen, Sekaran and Hale (1994) concentrated on the problem of achieving coordination without sharing information. As an illustration application, the block-pushing problem (here, two agents cooperate in moving a block to a specific predefined position) was chosen. The individual agents learn to jointly push the block toward its goal position. The learning task, hence, was divided among two agents. Sen and his colleagues showed that the joint learning goal can be achieved even if the agents do not model each other or exchange information about the problem domain. Instead, each agent implicitly takes into consideration the other agent's activities by sensing the actual block position. Insofar as the individual agents do not exchange information and learn according to the same algorithm (known as Q-learning), this work is also related to multiplied learning.

Plaza, Arcos and Martin (1997) applied multi-agent case-based learning in the domain of protein purification. The task to be solved here is to recommend appropriate chromatography techniques to purify proteins from tissues and cultures. Each agent is assumed to have its own case base, where a case consists of a protein–technique pair, as well as its own methods for case indexing, retrieval and adaptation. Two different modes of "case-based cooperation" are proposed: first, an agent's case base can be made accessible to other agents; second, an agent can make its methods for handling cases available to other agents. These modes are applied when a single agent is not able to solve a given protein purification task, and both result in the division of a single learning task among several agents.

Parker (1993) concentrated on the question of how cooperative behavior can be learnt in multi-robot environments. Parker was particularly interested in the question of how coherent cooperation can be achieved without excessive communication. The main idea was to realize cooperative behavior by learning about one's own and others' abilities. The scenario consisted of heterogeneous robots that have to complete some predefined tasks as fast as possible, where the individual robots differ in their abilities to carry out these tasks. Learning is divided insofar as the robots collectively learn which robot should carry out which task in order to improve cooperation. Insofar as this approach requires that the robots learn about each other (and, hence, to develop a model of the others) in order to achieve a shared goal, it is closely related to interactive learning. Other interesting research on multi-robot learning that falls into the category of divided learning is described in Mataric (1996).

Weiss (1995) dealt with the question of how several agents can form appropriate groups, where a group is considered as a set of agents performing compatible actions that are useful in achieving a common goal. As an application domain, the multi-agent blocks world environment was used. The task to be solved in this domain was to transform initial configurations of blocks into goal configurations, where the individual agents are specialized in different activities. The proposed algorithm provides mechanisms for the formation of new groups and the dissolution of old ones. Formation and dissolution are guided by the estimates of the usefulness of the individual agents and the groups, where the estimates are collectively learnt by the agents during the problem-solving process. Learning is divided insofar as different agents are responsible for different blocks and their movements. This work is related to interactive learning insofar the agents and groups dynamically interact (by building new groups or dissolving old ones) during the learning process whenever they detect a decrease in their performance.

## *3.3 Interaction Mechanisms*

### 3.3.1 General

The multiplication and division mechanisms do not explain all the benefits of multi-agent learning. Other learning mechanisms are based on the fact that agents interact *during* learning. Some interaction also occurs in the two previous mechanisms, but it mainly concerns the input or output of the individual agents' learning processes. Moreover, in the case of multiplied and divided learning, interaction typically occurs at the relatively simple and prespecified level of "pure" data exchange. In the case of interactive learning, the interaction is a more dynamic activity that concerns the intermediate steps of the learning process. Here, interaction does not just serve the purpose of data exchange, but typically is in the spirit of a "cooperative, negotiated search for a solution of the learning task". Interaction, hence, is an essential part and ingredient of the learning process. An agent involved in interactive learning does not so much act as a generalist or a specialist, but as a "*regulator*", influencing the learning path, and as an "*integrator*", synthesising the conflicting perspectives of the different agents involved in the learning process. For an illustration of the differences between the three classes of learning, consider the concept learning scenario. In an only-multiplied system, two inductive agents could interact to compare the concepts they have built. In an only-divided system, a "negative instances manager" could refute the concept proposed by the "positive instances manager". The resulting concept may be better but the reasoning of one agent is not directly affected by the others. In an interactive approach, two inducers could negotiate the generalization hierarchy that they use. The term "interaction" covers a wide category of mechanisms with different potential cognitive effects, such as explanation, negotiation, mutual regulation, and so forth. The complexity of these interactions makes up another difference between interactive learning on the one hand and multiplied/divided learning on the other.

The difference between the interactions in the multiplication/division approaches and the interaction approaches is mainly a matter of granularity: agents using the "multiplication/division" mechanisms interact about input and/or output of the agents" learning processes, while the mechanisms classified under the "interaction" label concern intermediate learning steps. Obviously, the difference between an input and an intermediate step is a matter of granularity, since intermediate data can be viewed as the input/output of sub-processes. The criterion is hence not fully operational. However, the interaction mechanisms (e.g. explanation) are also concerned with how the intermediate data are produced. We briefly illustrate this point in the context of case-based reasoning:

- *Multiplication*: case-based reasoning agents interact about the solution they have derived by analogy.
- *Division*: a "case-retrieval" agent asks a "case-adapter" agent for more-precise specification of the target case.
- *Interaction*: a "case-retrieval" agent asks a "case-adapter" agent for justification for the selection of a particular case or for naming a criterion that should be used for retrieval.

### 3.3.2 Examples of "Interactive Learning"

Sugawara and Lesser (1993) investigated how coordination plans can be learnt in distributed environments. The diagnosis of network traffic was chosen as an application domain. By observing the network, each agent responsible for diagnosing some part of the network learns by observation of a model of the network segment in which it is located and of the potential failures that may occur. Based on this learning, the agents collectively develop rules and procedures that allow improved coordinated behavior (e.g. by avoiding redundant activities) on the basis of diagnosis plans. The learning task is not divided *a priori*; instead, the agents dynamically contribute to the joint learning task depending on their actual knowledge.

Bui, Kieronska and Venkatesh (1996) studied mutual modeling of preferences as a method for reducing the need for time- and cost-consuming communication in negotiation-intensive application domains. The idea was to learn from observation to predict others' preferences and, with that, to be able to predict others' answers. The work is centered around the concept of joint intentions and around the question of how such intentions could be incrementally refined. This incremental refinement requires conflict resolution during a negotiation process. As an application domain, they chose a distributed meeting scheduling scenario in which the agents have different preferences with respect to the meeting date. Moreover, it is assumed that none of the agents possesses complete information about the others, and it is taken into consideration that the privacy of personal schedules should be preserved.

In the context of automated system design, Nagendra Prasad, Lesser and Lander (1995) investigated the problem of how agents embedded in a multi-agent environment can learn organizational roles. The problem to be solved by the agents is to assemble a (simplified) steam condenser. A role is considered to be a set of operators an agent can apply to a composite solution. Each agent is assumed to be able to work on several composite solutions concurrently, and to play different roles concurrently. Three types of organizational roles were distinguished: namely, for initiating a solution, for extending a solution, and for criticizing a solution. It was shown that learning of organizational roles enables a more efficient, asynchronous and distributed search through the space of possible solution paths. Although role learning itself is done by an agent independent of the other agents, all agents are involved in an overall conflict detection and resolution process. This work is also of interest from the point of view of "divided learning" insofar as the organizational roles are predefined and the agents are responsible for different parts of the steam condenser (e.g. the "pump agent" and the "shaft agent").

In closely related work, Nagendra Prasad, Lesser and Lander (1996) investigated multi-agent case-based learning in the context of automated system design. As in their aforementioned earlier work, the problem of building a steam condenser was chosen as an illustrative application. Here, the focus was not on organizational roles, but on negotiated case retrieval. Each individual agent is assumed to be responsible for integrating a particular component of the overall system, and to have its particular view of the design problem. Moreover, each agent is associated with its own local case base and with a set of constraints describing the component's relationships to other components, where a case consists of the partial description of a component configuration and the constraints associated with it. The agents negotiate and iteratively put together their local "sub-cases" such that constraint violations are reduced and a consistent overall case that solves the design problem is achieved. With that, both the cases and the processes of case indexing, retrieval and adaptation are distributed over several agents, and learning requires intensive interaction among the individuals.

Tan (1993) dealt with the question of how cooperation and learning in multi-agent systems are related to each other, where exchange of basic information about the environment and exchange of action selection policies are distinguished as different modes of cooperation. Experiments were conducted in the predator/prey domain. The results indicated that cooperation can considerably improve the learning result, even if it may slow down learning, especially at the beginning. This work is closely related to divided learning insofar as Tan also considered situations in which learning and sensing are carried out different agents, and it is related to multiplied learning insofar as Tan considered situations in which different predators learn completely independent of each other.

## 4 The Psychological Perspective: Multi-agent Learning versus Human–Human Collaborative Learning

This section characterizes available approaches to multi-agent learning from the perspective of human–human collaborative learning. As noted in Weiss (1996), DAI and ML use the term "multi-agent learning" for two different meanings. First, in its stronger meaning, this term refers only to situations in which the interaction among several agents aims at, and is required for, achieving a common learning goal. Second, in its weaker meaning, this term additionally refers to situations in which interaction is required for achieving different and agent-specific learning goals. For a brief illustration of these meanings, consider the multi-agent scenarios described in section 2. The transportation domain is a good example for multi-agent learning in its weaker meaning; here, the interacting agents (trucks or companies) pursue different learning goals: namely, to maximize their own profit at the cost of the other agents' profit. The manufacturing domain is a good example for multi-agent learning in its stronger meaning, because here all agents or machines try to learn a schedule that allows the manufacturing of products in minimal time and with minimal costs. Obviously, it is multi-agent learning in its stronger meaning that is more related to human–human collaborative learning, and which is therefore chosen as a basis for all considerations in this section.

The three types of learning mechanisms introduced in the previous section allow us to contrast multi-agent and single-agent learning, but are not appropriate for comparing multi-agent with human–human collaborative learning. In human groups, multiplied, divided and interactive learning mechanisms occur simultaneously. "Multiplication" mechanisms occur. For instance, in a synchronous written communication environment we observed several instances (Dillenbourg et al., 1997) of peers producing the same sentence simultaneously; i.e. they were conducting in parallel the same reasoning on the same data. The "division" mechanisms do occur in collaborative learning. We discriminate (see chapter 1, this volume) cooperative settings, in which the division of labour is regulated by more or less explicit rules, and collaborative learning, with no explicit division of labour. However, even in collaborative tasks, a spontaneous division of labour can be observed (Miyake, 1986), although the distribution of work is more flexible (it changes over time).

Some of the mechanisms that make collaborative learning effective (Dillenbourg & Schneider, 1995) relate to the first and second types of the learning mechanisms. However, the "deep secret" of collaborative learning seems to lie in the cognitive processes through which humans progressively build a *shared understanding*. This shared understanding relates to the third type of mechanisms. It does not occur at once, but progressively emerges in the course of dialogue through processes like conflict resolution, mutual regulation, explanation, justification, grounding, and so forth. Collaborative dialogues imply a sequence of episodes, some of them being referred to as "explanations", "justification", or whatever, and they are all finally instrumental in the construction of a shared solution. The cognitive processes implied by these interactions are the focus of current studies on human–human collaborative learning, and they have been largely ignored so far in the available studies on multi-agent learning. In the following, we will therefore concentrate in more detail on three of these processes—conflict resolution, mutual regulation, and explanation—in order to show what current approaches to multi-agent learning do *not* realize. Additionally, a closer look is taken at the importance of misunderstanding, which usually is considered by (D)AI researchers as something that should be avoided in (multi-)agent contexts, but which plays an important role for human–human collaborative learning. With that, and in contrast to the previous section, this section provides a *"negative" characterization* of multi-agent learning that allows us to make suggestions for future research on this kind of learning.

We are speaking of "(different types of) mechanisms" because this chapter concentrates on computational agents. It is clear, however, that human–human collaborative learning constitutes a whole process in which "multiplication", "division" and "interaction" correspond to different aspects of the same phenomena. Therefore, in focusing on the processes of conflict resolution, regulation, and explanation, we illustrate these three regards.

### *4.1 Conflict Resolution*

The notion of conflict, which is popular in DAI, is the basis of the socio-cognitive theory of human–human collaboration (Doise and Mugny, 1984). According to this theory, the benefits of collaborative learning are explained by the fact that two individuals will disagree at some point, that they will feel a social pressure to solve that conflict, and that the resolution of this conflict may lead one or both of them to change their viewpoint. This can be understood from the "*multiplication*" perspective, because a conflict between two or several agents is originated by the multiplicity of knowledge. It appears, however, that learning is not initiated and generated by the conflict itself but by its resolution—i.e. by the justifications, explanations, rephrasements, and so forth, that lead to a jointly accepted proposition. Sometimes, a slight misunderstanding may be enough to generate these explanations, verbalizations, and so forth. Hence, the real effectiveness of conflict has to be modeled through interactive mechanisms.

### *4.2 Mutual Regulation*

Collaboration sometimes leads to improved self-regulatory skills (Blaye, 1988). One interpretation of these results is that strategic decisions are precisely those for which there is a high probability of disagreement between partners. Meta-knowledge is verbalized during conflict resolution and, hence, progressively internalized by the partners. Another interpretation attributes this effect to a reduced cognitive load: when one agent looks after the detailed task-level aspects, the other can devote his cognitive resources to the meta-cognitive level. This interpretation could be modeled through a "*division*" mechanism. However, mutual regulation additionally requires that each partner maintains "some" representation of what the other knows and understands about the task. This is not necessarily a detailed representation of all the knowledge and viewpoints of a partner, but a task-specific differential representation ("With respect to that point, my partner does have a different opinion to me"). The verbal interactions through which one maintains a representation of one's partner's knowledge have been partly studied in linguistics, but the cognitive effects of this mutual modeling process constitute an interesting item on the agenda of psychology research. Interestingly, this ability was proposed by Shoham (1990) as a criterion to define what an agent is and whether an agent is more than a simple function in a system.

### *4.3 Explanation*

The third phenomenon considered here is learning by building an explanation to somebody else. Webb (1989) observed that subjects who provide elaborated explanations learn more during collaborative learning than those who provide straightforward explanations. Learning by explaining does even occur when the learner is alone: students who explain aloud worked-out examples or textbooks, spontaneously or on a experimenter's request, acquire new knowledge (Chi et al., 1989). The effect of explanation in collaborative learning could be reduced to a multiple self-explanation effect (see chapter 6, this volume). For instance, in a multi-agent system, each agent could be given a "learning by explaining" algorithm inspired by VanLehn's computational model of the self-explanation effect (VanLehn et al., 1992) or by some explanation-based learning algorithm (EBL) (Mitchell et al., 1986). The benefit of "*multiplicity*" could result from the heterogeneity of outcomes: the generalization stage—the most delicate step in EBL—could be "naturally" controlled as the least general abstraction of the proofs built by different agents.

An explainer has to simultaneously build the explanation and check for its consistency. In self-explanation, this self-regulation process is costly, while in collaboration, the cost of regulation can be shared. The cognitive load is lower when one checks the consistency of an explanation given by somebody else than when one has to check one's own explanation. Hence, this interpretation mechanism can be viewed from the perspective of the second type of mechanisms, the division of labor. This idea was implemented in "people power" (Dillenbourg & Self, 1992). An agent explains a decision by showing the trace of the rules fired to take that decision. Some of these rules were too general and learning consisted in specializing them progressively. When the second agent received an explanation, he tried to refute the argument made by the first agent (in order to show that it was too general). Actually, both agents used exactly the same rules—hence, learning was not related to multiplicity. It simply was too expensive for one agent to build a proof, to try to refute any step of its proof, to refute any step of this self-refutation, etc. In other words, the functions "produce an explanation" and "verify an explanation" were distributed over two agents, as in the "*division*" perspective.

However, explanation is most relevant from the perspective of the *interaction* mechanisms. An explanation is not simply something that is generated and delivered by one agent to another (as it could be viewed in the case of multiple self-explanation). Explanation is, of course, interactive. Current studies on explanation view it as a mutual process: explanation results from a common effort to understand mutually (Baker, 1992). The precise role of interactivity is not yet very clear. It has not been shown that interactive explanations generate higher cognitive effects. The situations in which the explainee may interact with the explainer do not lead to higher learning outcomes than those where the explainee is neutral (see chapter 6, this volume).

### 4.4 The Importance of Misunderstanding

The three above-mentioned mechanisms are modeled in some way in DAI. The main difference is that the computational models use formal languages while humans use natural language (and non-verbal communication). It is, of course, common sense to use a formal language between artificial agents, but the use of natural language creates a major difference that might be investigated in DAI without having to "buy" completely the natural language idea: it creates *room for misunderstanding*.

Misunderstanding leads to rephrasement, explanation, and so forth, and hence to processes that push collaborative learners to deepen their own understanding of the words they use. As Schwartz (1995) mentioned, it is not the shared understanding *per se* that is important, but the efforts toward shared understanding. We see two basic conditions for the emergence of misunderstanding:

1. Misunderstanding relies on semantic ambiguity. It is a natural concern of multi-agent system designers to specify communication protocols in which the same symbol conveys the same "meaning" for each agent, simply because "misunderstandings" are detrimental to system efficiency. With that, it turns out that there is a conflict between efficiency and learning in multi-agent systems: misunderstanding and semantic ambiguity increase the computational costs, but, at the same time, the potential of collaborative learning.
2. Having room for misunderstanding also means having room for discrepancies in the knowledge and experience of interacting individuals. Such discrepancies may arise, for instance, because the individuals act in different (perhaps geographically distributed) environments, because they are equipped with different sensing abilities (although they might occupy the same environment), and because they use different reasoning and inference mechanisms. From this point of view, it turns out that heterogeneity should be not only considered as a problem, but also as an opportunity for designing sophisticated multi-agent systems.

This room for misunderstanding implies that agents are (partly) *able to detect and repair misunderstanding*. Obviously, it is not sufficient to create space for misunderstanding, but the agents must be also capable of handling it. Any misunderstanding is a failure in itself, and it constitutes a learning opportunity only if it is noticed and corrected. The mechanisms for monitoring and repairing misunderstandings in human contexts are described extensively in this volume (chapter 3, this volume). What matters here, from the more technical multi-agent perspective, is that the monitoring and repairing of misunderstandings require the agent to maintain a representation of what it believes and what its partners believe (see section 4.2).

These considerations seem to indicate that multi-agent learning necessarily has to be limited with regard to collaboration, as long as the phenomenon of misunderstanding is suppressed and ignored in multi-agent systems. A solution to this limitation would be to consider "miscommunication" not always as an error that has to be avoided during design, but sometimes as an opportunity that agents may exploit for learning. The same distinction can be made with linguistic studies, where the concept of "least collaborative effort" is used to describe how interlocutors achieve mutual understanding, while we prefer to refer to the "optimal collaborative effort" to emphasize that the additional cognitive processes involved in repairing mutual understanding may constitute a source of learning (Traum & Dillenbourg, 1996).

Two additional points should be made here. First, the above considerations may sound inappropriate since the notion of (mis)understanding is purely metaphorical when talking about artificial agents. However, one can understand the notion of "meaning" as the relationship between a symbol and a reference set. Typically, for an inductive learning agent, A1, a symbol X will be related to a set of examples (S1). Another agent, A2, may connect X with another set of examples (S2), overlapping only partly with S1. Then, negotiation can actually occur between A1 and A2 as a structure of proofs and refutations regarding the differences between S1 and S2.

Second, in addressing the above question, we restricted ourselves to verbal interaction and dialogues, because this is most interesting from the point of view of artificial agents. In particular, we did not deal with non-verbal interaction (based on, for example, facial expressions and body language), even though it is an important factor in human–human conversation. However, as a first step in that direction, we are currently working on implicit negotiation in virtual spaces where artificial agents negotiate division of labor by movements: agent A watches which rooms are being investigated by agent B and acknowledges agent B's decision by selecting rooms that satisfy the inferred strategy (Dillenbourg et al., 1997). Finally, it should be noted that we did not deal with certain aspects of dialogues that are relevant to human–human collaborative learning, but are difficult to model in artificial contexts. Examples of such aspects are the relationship between verbalization and consciousness (humans sometimes only become aware of something when they articulate it), dialogic strategies (one purposefully says something that one does not believe to test one's partner) and the internalization of concepts conveyed in dialogues (these points are addressed in chapter 7, this volume).

# 5 Conclusions

## *5.1 Summary*

This chapter aimed at taking the first steps toward answering the question "*What is "multi" in multi-agent learning?*" We attacked this question from both an ML and a psychological perspective. The ML perspective led to a "positive" characterization of multi-agent learning. Three types of learning mechanisms—multiplication, division and interaction—that can occur in multi-agent but not in single-agent systems were distinguished. This shows that multi-agent learning can take place at a qualitatively different level compared with single-agent learning as it has been traditionally studied in ML. Hence, multi-agent learning is more than just a simple magnification of single-agent learning. The psychological perspective led to a "negative" characterization of multi-agent learning. Several processes like conflict resolution, mutual regulation and explanation were identified that play a more significant role in human–human collaborative learning than in multi-agent learning, since they contribute to the elaboration of a shared understanding. The cognitive effort necessary to build this shared understanding, i.e. continuously to detect and repair misunderstanding, has not received enough attention in multi-agent learning where noise is *a priori* not treated as a desirable phenomenon. Hence, despite the fact that multi-agent learning and human–human collaborative learning constitute corresponding forms of learning in technical and in human systems, it is obvious that the available multi-agent learning approaches are of lower complexity.

## *5.2 Some Implications*

There are many approaches to multi-agent learning that are best characterized as multiplication or division mechanisms, but fewer that are best characterized as interaction mechanisms. This indicates important potential directions for future research on multi-agent learning. What is needed are algorithms according to which several agents can interactively learn and, at the same time, influence—trigger, redirect, accelerate, etc.—each other in their learning. From what is known about human–human collaborative learning, the development of these kinds of multi-agent algorithms may be facilitated by putting more emphasis on the implementation of negotiation processes among agents, triggered by the possibility of misunderstanding and oriented toward the emergence of a shared understanding.

# Acknowledgments

We are grateful to the European Science Foundation (ESF) programme "Learning in Humans and Machines" for the opportunity to work together in the past years. This chapter summarizes the results of our co-work.

# 5

# Comparing Human–Human and Robot–Robot Interactions

Richard Joiner, Kim Issroff and John Demiris

In this chapter, we compare current approaches to analysing multi-robot systems with those that analyse human–human collaborations. The chapter reviews recent research concerning the analysis of multi-robot systems and research on human–human computer-supported collaborative learning. It then compares the two areas, finding important differences between them that influence the type and nature of research conducted. In the second part of the chapter, we discuss a study ($n = 18$) that was carried out to investigate humans solving a problem using the task, methods and means of evaluation used in multi-robot research. The participants had to collect five flowers using two bees. Each adult was controlling one of the bees. Techniques from both distributed artificial intelligence and computer-supported collaborative learning were used to investigate the nature of their interactions. The analysis showed that the methods used in multi-robot systems can be applied to the analysis of human–human interactions and this has important implications for both.

## 1 Introduction

With the flourishing of research on human–human collaborations and research on robot collaborations, this chapter is a comparison of the analysis approaches that researchers in their respective fields use to investigate collaborations. These two areas have taken quite different approaches to their investigations and this chapter aims to present a comparison of these approaches. The scope of the chapter is limited to discussing multi-robot systems rather than including multi-agent systems, and the discussion of human agents collaborating using computers is limited to groups of children working around a computer rather than computer-mediated interactions. We then compare the two areas and find that they have fundamentally different motivations. These different motivations result in differences in methods, analytic schemes and tasks. Applying the tasks, methods and means of evaluation of one area to another could have implications for both. A case study of human–human collaborations is presented that uses a typical multi-robot system task (foraging) and a well-established multi-robot analysis method, trace diagrams. This is discussed in terms of its usefulness for understanding human–human collaborations and robot–robot collaborations.

Both fields have no accepted definitions of collaboration or cooperation. Papers on computer-supported collaborative learning in humans generally assume that collaboration is occurring when a group of individuals work together on a task using a computer. There is a parallel situation with respect to analysing robot interactions, in that robots are seen to collaborate or cooperate when they are given a task to achieve and programmed to behave in ways perceived by the programmer to be collaborative.

Research on cooperating robots started in the early 1980s, but was initially concerned mainly with the fundamental issues involved in coordinating multiple robots (initially robot manipulators) in order to increase the productivity of robotic systems. During the 1980s, the benefits of using multiple robots were increasingly realised and issues such as task decomposition, communication and learning became increasingly important. However, it was not until the early 1990s that the robots started to be used as "tools" specifically to study such issues along with issues like group behaviour, the effects of communication on group performance, and group learning.

Research on the ways in which humans learn collaboratively when using computers originated in the late 1980s with an initial emphasis on experimental methodologies driven by psychologists trying to understand aspects of cognition. With the advent of groupware technology to support human collaborations, there has been a shift to ecologically valid research settings with an emphasis on learning. This shift has been paralleled by a shift in methodologies that have moved from concentrating on outcomes to investigating the process and nature of interactions while people work together.

## 2 Multi-robot Systems

During the last few years, there has been an increased interest in the study of multiple robot systems. In this section, we describe the motivations behind such research, the type of experiments performed along with the parameters studied, and the methods of analysis used for the interpretation of the results.

### *2.1 Motivations*

There are a number of engineering reasons for building multi-robot systems. There might be cases where building a large collection of simple robots is actually cheaper than building a complex single-robot solution to a problem. For example, mass-produced simple autonomous robots have been proposed as an alternative to current single-robot attempts at planet exploration (Brooks & Flynn, 1989). The resulting systems will also be more fault-tolerant, which means that even if a percentage of these robots fail, the collection will still be able to complete the task. Also, there could be tasks that a single robot cannot complete efficiently or cannot complete at all. For example, Paljung and Yun (1993), describe experiments they did with a dual robot-arm system manipulating large objects that none of the robot arms could handle by itself.

There are also a number of scientific reasons for building multi-robot systems. They can be used as a tool to model interactions between agents within a group. For example, there has been considerable research modelling the interactions between ant-type robots, and showing how simple interactions can result into complex group behaviours (Deneubourg et al., 1991). Most of this research has been performed in simulation, since it is difficult to design, build, program and maintain a large collection of robots (Huber and Kenny, 1994).

It is important to understand the motivations behind research concerned with multiple robot systems since the tasks, analysis and evaluation of such systems is largely dependent on the motivation. This dependence is evident from the choice of evaluation metrics (section 2.3) that are used by researchers in this discipline. In the following sections, the kinds of parameters varied in a multi-robot system are described and the procedures used to evaluate them are discussed.

## *2.2 Range of Parameters Studied*

There are at least five parameters that can be modified in a multi-robot system: the nature of the interaction between the individual robots, the type of communication used between them, the task in hand, the complexity of the environment, and the capabilities of the individual robots.

### 2.2.1 Nature of Interaction between Individual Robots

Robots can be designed to interact with each other in a number of ways. They can be built to compete for resources between themselves. They can also be designed to work cooperatively by not interfering with each other. Social laws are usually applied to resolve any conflicts (e.g. traffic laws when their paths meet). They can also be designed to work collaboratively by inferring the plans and goals of the other robots and attempting to assist when/if they can (e.g. in pushing or manipulating an object that cannot be handled by a single robot).

### 2.2.2 Communication

There are essentially three types of communication used in multi-robot systems: communication through the environment (e.g. if a robot moves a box in the appropriate place, another robot will see it, pick it up and move it); implicit communication via observation/sensing (e.g. when a robot attempts to move a box that it is not capable of moving alone, a second robot might detect this and move to assist); and explicit communication where messages are intentionally sent from one robot to others. The content of such explicit messages vary from information about the position of a robot, to information about the particular state of a robot, and to information concerning the robot's goal and/or plans.

### 2.2.3 The Task

To avoid the danger of designing a multi-robots system that is only optimised for a specific task, researchers usually test their systems on a variety of tasks. Workers using real mobile robots employ tasks that involve searching a predefined space, moving in a specific formation, and manipulating or moving objects. Those who use simulations employ a greater variety of tasks and this eliminates most of the technical problems of dealing with real robot hardware.

#### 2.2.4 The Complexity of the Environment

Researchers investigate the performance of a multi-robot systems by varying the environment in which the multi-robot system is operating. The task can be varied in a number of ways: by altering the number of objects to be collected, and the number of obstacles, by changing the resources available and by altering the dimensions of the search space.

#### 2.2.5 Capabilities of Individual Robots

Although some of parameters of the robot will be determined by the task and the complexity of the environment, researchers also vary the physical, sensor, and cognitive abilities of the robots. The resulting system can have either a homogeneous or a heterogeneous set of robots. The capabilities of the robots can be modified so that some of them have different roles (leaders/planners vs. specialised workers).

### *2.3 Evaluation*

Evaluation methodologies in multi-robot systems have begun to move away from abstract claims about "better performance" and "more robust systems" to the development of specific metrics that could be used to justify such claims. A variety of engineering metrics are sometimes used, such as the cost of building the system, or the energy required to complete a task, but these are not within the scope of this chapter. Although, currently, there is not an agreed set of standard metrics, there are a number that are commonly used, such as time, reliability/repeatability, scalability, optimality, and fidelity.

#### 2.3.1 Time

One of the most commonly used performance metrics is the time that is needed for a multi-robot system to complete a task. A slight variation used is the number of iterations, which is more commonly used in simulations to avoid the problem of comparing experiments carried out on different computers.

#### 2.3.2 Reliability/Repeatability

Another common performance metric is the reliability of the system. Experiments are usually repeated several times to establish the reliability of the findings.

### 2.3.3 Scalability

A third performance metric is the scalability of the system. Researchers investigate whether the performance characteristics remain the same if the scale of the problem is changed (e.g. changing the number of robots, the number of obstacles or the dimensions of the search space). The performance of the system is degraded when the problem is made more difficult. Researchers then examine the nature of this degradation by investigating whether it degrades gracefully or catastrophically.

### 2.3.4 Optimality

A fourth performance metric is the optimality of the solution. Often, tasks can be solved in a number of different ways and the solution found by the system is compared with the optimal solution.

### 2.3.5 Fidelity

Fidelity is the fifth performance metric used in the evaluation of multi-robot systems. Since one of the motivations for building multi-robot systems is to model natural systems, they are frequently compared with the behaviour of these natural systems. A variety of techniques exist for assisting in the collection and analysis of multiple robot data. Since most robot tasks involve movement in space, data often come in the form of visual traces. These traces usually assist in understanding the behaviour of the system and identification of areas for improvement (e.g. particular areas where interference is high, or whether the agreed traffic laws have the desired effects, etc.). Such traces often assist the researcher in examining how well the behaviour of the system resembles the behaviour of the corresponding natural system.

Performance metrics are usually plotted against the different parameters that were described above. However, if learning is involved, the performance of multi-robot systems is plotted over time to determine whether the system eventually learns. In section 2.4, some common tasks are described where these metrics have been applied.

## *2.4 Example Tasks*

One of the most common tasks used in multiple robot experiments is the foraging task, mainly because it is a task that is clearly defined and offers easily measurable evaluation criteria. It is inspired by biological systems, such as those of ants and various other animals, so it offers potential for comparing the artificial solutions with the natural ones. In its robotic form, the foraging task is defined as follows: we are given a (usually enclosed) search area, a number of objects (the "food"—usually pucks or other small, light objects), a number of robots and a home region. The researcher programs the robots in such a way that they will be able to find efficiently as many objects as possible, returning them to the home region. Although the robots could be programmed through a central system that would divide the task, assign different search areas to different robots and monitor the execution, most researchers prefer to program each robot independently and study the effects of communication and different individual search strategies on group performance.

Initially, the robots are placed in the home region and the objects are scattered around the search area. The robots are then powered on and allowed to move in the search area. When they locate an object they grasp it, bring it to the home region and return to the search area to find more. A camera is placed on top of the search area to obtain visual traces of the robot movements, or alternatively, the robots communicate their position to a computer that gathers these data and plots the traces later. When all the objects have been collected or a specific time limit has been reached, the robots are stopped, and the data are analysed according the metrics described earlier—usually, the visual traces reveal the amount of interference between the robots and the overlap between their searched areas.

The experiments are then repeated, modifying the amount and the type of communication between the robots. Interesting results from robot research using the foraging task include the following:

- Simple communication (such as transmitting information about each robot's state to each of the others) improves the performance significantly (Balch & Arkin, 1994).
- More complex communication strategies (such as communicating each robot's goal to the other robots) offers little benefit over basic communication (such as communicating only your state information) for this task (Balch & Arkin, 1994).
- Cooperation emerges very quickly in small populations of robots, but it is counterbalanced in bigger populations by the competition between the robots (Drogoul & Ferber, 1992).

Foraging comes in other forms. Consumption is one variation of the foraging task. In this task, robots consume the food on the spot and continue searching rather than returning home with the food as in foraging. Balch and Arkin (1994) conducted some experiments where they varied the communication strategy, and evaluated whether this affected the performance of the system. They reported similar findings to those found with the foraging task. Communication improved performance compared with no communication, but more complex communication did not offer significant benefit over simple communication.

Searching is another variation of the foraging task and also has a number of different varieties, such as grazing (Balch & Arkin, 1994) and janitorial service (Parker, 1992). In searching tasks, performance is often evaluated by assessing the percentage of the space covered by the robots. Parker (1992) built and evaluated a number of multi-robot systems that were designed to find an adaptive architecture that would allocate sub-tasks to the different robots. Balch and Arkin (1994) compared different communication strategies and demonstrated, unlike their previous studies with foraging and consuming, that explicit communication does not improve the performance of multi-robot systems. They argue that this finding was because implicit communication through the environment already exists, as areas that have been grazed can be detected by other robots and avoided.

## 3 Computer-Supported Collaborative Learning

There is now a considerable body of research on how people work in groups when they are using computers (see, e.g. Crook, 1994; Foot et al., 1994; Koschmann, 1996). The way in which researchers have studied these types of collaborations has evolved over several decades and contributions have been made from the fields upon which educational technology is founded: psychology, education and computer science. A detailed discussion of the ways in which human–human collaborations are studied is beyond the scope of this chapter. We will first outline the motivation for studying these types of interactions, detail the range of factors manipulated and finally discuss the ways in which human–human collaborations are evaluated.

### *3.1 Motivations*

There are three main reasons for studying computer-supported collaborative learning. One is that a number of theories of cognitive development propose that social interaction is central for development, and research into computer-based collaborative learning has been used to investigate those theories. For example, in Piagetian theory the resolution of interindividual conflicts is thought to facilitate cognitive development. This theory has been tested by examining the relationship between the number of conflicts in computer-supported interactions and learning outcome (Clements & Nastasi, 1988; Nastasi & Clements, 1992).

Another reason for studying computer-supported collaborative learning is to provide education guidelines for the optimal use of this type of group activity, because, due to resource limitations, most activity around a computer in schools occurs in groups. For instance, Trowbridge (1987) has tried to identify what is the optimal group size for collaborative group work around a computer.

The third reason for research into computer-supported collaborative learning is to inform the design of computer systems designed to support it. A number of computer tools have been designed for supporting collaborative work and, recently, researchers have looked at how they can be used in supporting collaborative learning. For instance, Koschmann et al. (1996) have developed a collaborative learning environment for supporting problem-based learning for preclinical medical education. In this environment, participants have both shared and personal workspaces and can communicate with other participants using an electronic chat system.

### 3.2 Range of Factors Studied

For the above reasons, researchers have investigated factors that are thought to be important for collaborative learning.

#### 3.2.1 Nature of the Interaction between Students

There have been numerous studies that have compared the performance of people working individually versus those working in pairs. For example, Blaye et al. (1991) found that pairs performed better than individuals on a computer-based problem and that this beneficial effect transferred to later individual work. Similar beneficial effects have been reported by Amigue (1989), Amigue and Agostinelli (1992), Mevarech, Stern and Levita (1987) and Mevarech (1993). However, Jackson and Kutnick (1996) report a study on the effects of peer interaction that found that individuals performed better than pairs. They conclude that the beneficial effects of peer interaction are dependent on the nature of the task.

Others have investigated the types of group interaction and how that affects learning. One set of studies have compared people working competitively with those working cooperatively. For example, Johnson, Johnson and Stanne (1985) and Johnson and Johnson (1986) found that children who worked cooperatively were more successful in solving a complex computer-based problem than children who worked competitively or individually and that they performed better on subsequent individual post-tests. There have also been a number of observational studies that have examined the types of interaction that go on between children. Pozzi, Healey and Hoyles (1993) and Healey, Pozzi and Hoyles (1995) categorised groups of children into three different styles of organisation: integrated, connected and fragmented. They found that for some tasks, a fragmented style was optimal, while for other tasks, a connected or integrated style was best.

#### 3.2.2 Communication

Several studies have examined whether communication is important in learning by comparing groups whose members are allowed to communicate with each other with groups whose members are not. There have been mixed results, with several studies showing that if groups are not allowed to talk, they do as well as those groups who are allowed to talk (Jackson, Fletcher & Messer, 1992), whereas others have found that the groups allowed to communicate performed better than those who did not (Teasley, 1995). Other researchers have investigated the role of talk in collaborative learning by examining the relationship between different types of talk and learning outcome. Aspects of talk that have been examined are on- and off-task talk (Bennett & Dunne, 1991); social- and cognitive-oriented talk (Wild & Braid, 1996); the level of help given and received, and the appropriateness of responses to requests for help (Webb, 1984); interindividual conflicts (Clements & Nastasi, 1988; Nastasi, Clements & Battista, 1990; Nastasi & Clements, 1992); and negotiation and co-construction (Barbieri & Light, 1992).

### 3.2.3 The Task

Other researchers have studied the ways in which the task affects both the nature and effectiveness of collaborative interactions. For example, Crook (1987) found that certain tasks lend themselves more to collaboration than others. Nastasi and Clements compared pairs of children working with LOGO with pairs of children working either with computer-assisted instruction (Clements & Nastasi, 1988; Nastasi, Clements & Battista, 1990) or computer-based writing (Nastasi & Clements, 1992). They found in both studies that pairs of children working in the LOGO environment were exhibiting more productive behaviour that fostered cognitive growth than pairs in the other environments.

### 3.2.4 Computer Environment

The computer environment has been used in a variety of ways for supporting collaborative learning. Koschmann (1996) identifies five such uses. One is as a means of simulating a problem for study to help to situate it in a real-world context. A good example of this is the Jasper series Cognition and Technology Group at Vanderbilt (1992) where, with the use of video-based, narrative adventures, students have to tackle complex problems in a meaningful context. Another use of technology is to mediate communication. Riel (1996) investigated the use of electronic networks for support within classroom and cross-classroom project work. A third is for introducing new resources. Pea (1996) developed a computer simulation where students could investigate geometrical optics. The simulation provided them with a resource that they could use to test and develop their ideas about light, reflectors and absorbers—in a way that they could not have done previously. A fourth is for archival storage. Scardamalia and Bereiters (1996) use technology as a means of providing a shared database where students can jointly construct a knowledge base in a manner similar to the development of scientific knowledge. The fifth use of technology is as a representational formalism. In a study of collaborative learning, Roschelle and Teaseley (1995) used a direct manipulation graphical simulation of the concepts of velocity and acceleration that explicitly represented those concepts on the computer.

Another way that computers have been used for supporting collaborative learning is as a learning companion. Several researchers have developed systems that simulate a virtual learning companion to interact with the student in ways that are thought to promote learning (Dillènbourg & Self, 1992; Chan & Chou, 1995). Finally, others have used the computer as a tutor to support collaborative learning (Chan & Baskin, 1990).

### 3.2.5 Qualities of Groups of Students

Recently, there has been a considerable amount of interest looking at mixed-gender and single-gender groups. Hughes et al. (1988) found that girls benefited from working with boys, but most of the other research has found no difference (Barbieri & Light, 1992) or that girls are at a disadvantage in mixed-gender pairs (Siann &

McLeod, 1986; Siann et al. 1988; Dalton, 1990; Underwood, McCafferey & Underwood, 1990; Lee, 1993; Underwood, Underwood & Turner, 1993; Underwood, Jindhal & Underwood, 1994). Recent studies by Yelland (1993, 1994, 1995) provide some explanation for this discrepancy in findings. She found that differences in performance due to the gender composition of the pair were dependent on the type of task and the time frame of the research design. A similar finding was reported by Underwood and Underwood (1997).

Other researchers, drawing on Piagetian notions of cognitive conflict and Vygotskian notions of the zone of proximal development, have conducted studies comparing groups of students with similar cognitive conceptions to those with different conceptions. For example, Howe et al. (1992) and Howe, Tolmie and MacKenzie (1995) used a computer-based physics problems to compare pairs of students with differing abilities with pairs of students with the same ability and found that student in pairs with differing abilities benefited the most.

### *3.3 Evaluation*

The factors discussed in section 3.2 have been systematically manipulated in research. Learning and socio-affective outcomes have been used to evaluate the effectiveness of collaboration. Individuals are assessed on either one or several measures before and after the intervention. Any changes in the measures are then attributed to the intervention. Socio-affective measures are predominantly used by educational sociologists who assume that collaboration can facilitate interpersonal relationships, which, in turn, has a positive effects on students' motivation and self-esteem (Johnson, Johnson & Maruyama, 1983). Nastasi and Clements (1991) reviewed studies on collaborative learning that have shown that it has a beneficial effect on motivation, attitudes towards learning, self-esteem and self-perception. Learning outcome is predominately used in research on collaborative learning from a cognitive perspective. Studies from this perspective have been conducted to evaluate the effectiveness of computer-supported collaborative learning on children's understanding of physics (Howe et al., 1992), computer programming (Nastasi & Clements, 1992), electronics (Amigue, 1989) and a whole range of educationally applicable tasks.

## 4 Comparing Research into Multi-robot Systems with Research into Computer-Supported Collaborative Learning

There are at least four ways in which research in these two areas differ. These differences shape the nature and type of the research conducted and are primarily due to differences in motivation and differences between humans and robots.

### *4.1 Motivations*

The motivation for building multi-robot systems is to ensure that robots perform a particular task in the most efficient way or to model natural systems. The motivation for studying collaborative learning is to inform the design of environments for supporting collaborative learning. The different motivations have a significant impact on the research that is carried out in the respective disciplines. These are discussed in more detail in the following.

### *4.2 Tasks*

The difference in motivation between the two research areas leads to differences in the type of tasks employed. Research into multi-robot systems uses tasks that have a practical engineering purpose. The development of multi-robot systems to undertake foraging tasks has the aim of developing systems to explore environments where humans would be unable to exist (i.e. other planets). They also use tasks that have a biological parallel: e.g. the development of multi-robot systems to model the behaviour of a colony of ants. This contrasts with the type of tasks used in research on computer-supported collaborative learning, which are designed to be either educationally relevant or to investigate particular cognitive processes. For example, there is considerable research going on to investigate collaborative learning in science, mathematics and computers, for obvious educational reasons.

### *4.3 Evaluation*

A third difference between the two areas is in how they evaluate effective collaborative activity and this, again, is largely determined by their different motivations. Research into computer-supported collaborative learning and multi-robot systems defines effectiveness in different ways. For multi-robot systems, a number of engineering and biological criteria are used to measure the performance of the system. Engineering criteria include time to complete the task, reliability of the system and scalability. Biological criteria include the degree to which the system corresponds to biological data. Research regarding computer-supported collaborative learning evaluates the effectiveness of collaborative learning in educational terms by measuring socio-motivational outcomes or learning outcomes.

### *4.4 Robots and Humans*

The differences between robots and humans also has a major influence on the methods and research carried out. Robots have limited communication abilities. In contrast, even the youngest students used in studies of human–human interactions can communicate with one another. The nature of this communication provides researchers with one way of representing the nature of the interactions and also allows them, arguably, some insight into the knowledge, thoughts and feelings of the students involved. However, communication in robots is not as complex, nor is it used as a means to access the knowledge of the robots. The approach used by researchers using robots is a bottom-up one in which the robots are given as much communication ability as is necessary in order to complete the task, or the type and richness of communication is altered to observe its effect on group behaviour and performance. These differences in both the quality and quantity of communication have implications for the respective research methodologies. Currently, much of the research on human–human collaborations is focusing on the nature of the talk and the development of shared understandings (Mercer, 1995), whereas research in multi-robot systems is frequently investigating the extent to which collaboration can occur with little or no communicative abilities.

Another major difference between robots and humans is that researchers working with robots have direct access to the internal representations of robots, which enables them to understand what the robots are "thinking" at any particular point in time. However, students' representations are not directly accessible, which means that researchers cannot objectively know what the student knows, is thinking or feeling. There are a number of techniques for trying to infer the thought processes. They range from clinical interviews after the interaction to protocol analysis. These techniques have advantages and disadvantages, but neither method provides a complete access to the thought processes of humans. This means that researchers can provide only an interpretation of the thought processes of the human participants, whereas researchers studying multi-robot interactions have direct access to the robots' internal representations and processes.

## 5 Using a Common Multi-robot Task with Adults

To apply the tasks, the methods and the means of evaluation of one research area to the other area could change the type of results and could have implications for both. At present, it is not possible to use robots in a typical study of human–human collaboration because of their limited communication capabilities. However, it is possible to use the task, methods and means of evaluation typical of a multi-robot study with humans. This section reports a study of adults collaboratively completing a common multi-robot foraging task, and uses an analysis method, trace diagrams, which is normally used by researchers interested in multi-robot systems, to analyse the nature of the collaborative interactions between humans.

The study had three aims. The first was to investigate how adults would work together on a joint problem-solving task not used before in research into computer-supported collaborative learning. The second was to analyse their joint problem-solving with trace diagrams—a method of analysis used in multi-robot systems but not used before in computer-supported collaborative learning. The third was to analyse the type of communication between the participants, looking at when they communicated and what was the purpose of their communication. This follows from research in the robot literature (Balch & Arkin, 1994), where it was found that communication about the robots' location did affect performance but communication about the robots' goals did not.

### *5.1 Method*

#### 5.5.1 Participants

We evaluated the collaborative problem solving of 20 psychology students, aged between 21 and 25 years old, from the Institute of Psychology, University of Aarhus, Denmark. They were all volunteers.

#### 5.1.2 Procedure

The participants solved the task in pairs and sat in front of a computer in separate rooms. They were told they would be playing a collaborative game. Each experimenter demonstrated to their participant how to use the software and gave a brief, scripted introduction to the task. The participants then played a practice version of the game with some assistance from the experimenters. After this, the participants had the opportunity to ask questions and were again informed about the goal. They were also told to solve the game as quickly as possible.

#### 5.1.3 Task

The task, which was especially devised for the experiment, was a computer-based foraging task called the "Bumble Bees".

The task was set in a 5 × 5 grid (see Figure 1). The goal was to help the bumble bees collect five flowers and transport them home. The bumble bees can only carry one flower at a time. There were two bumble bees (Bumble and Bee); each participant controlled one of them. At the start of the game, both bees were at "Home" in square 13 and the flowers were positioned in squares 2, 4, 4, 6 and 21. The minimum number of moves needed to solve the problem is 32.

| | | | | |
|---|---|---|---|---|
| 1 | 2 | 3 | 4 | 5 |
| 6 | 7 | 8 | 9 | 10 |
| 11 | 12 | 13 | 14 | 15 |
| 16 | 17 | 18 | 19 | 20 |
| 21 | 22 | 23 | 24 | 25 |

Figure 1: Bumble Bee 5 × 5 grid

### 5.1.4 Computer Environment

The Bumble Bees was built especially for this study in a Lambda MOO environment and has a text-based interface. The participants also were given a paper map of the world (identical to Figure 1) to help them. The following is an example used to illustrate how participants communicated and carried out the task. Participants will only "“see” their partner when they are both in the same room together. The participants could communicate with each other by using the page command. In the example below, Bumble pages Bee and suggests that she searches rooms 14–25 and Bee searches rooms 1–12. Bee acknowledges by paging "“OK” to Bumble.

bee_room_type: R13
A large field
Bumble is here.
Obvious Exits: West (to R12), East (to R14), South (to R18), and North (to R8).
You sense that Bumble is looking for you in bee_room_type R13
She pages, “I'll search rooms 14-25”
**Page Bumble OK**
She receives your page

The participants could move from room to room by typing north, east, south or west. In the example below, Bee types "“north” and moves from room 13 to room 8. She then types “north” again and moves to room 3, and then types east and moves to room 4, where she finds two flowers.

**north**
bee_room_type: R8
A large field
Obvious Exits: West (to R7), East (to R9), South (to R13), and North (to R3).
**north**
bee_room_type: R3
A large field
Obvious Exits: West (to R2), East (to R4), and South (to R8).
**east**
bee_room_type: R4
A large field
You see flower 2 and flower 5 here.
Obvious Exits: West (to R3), East (to R5), and South (to R9).
**Page Bumble I've found two flowers in room 4**
She receives your page
**take flower 2**
You take flower 2.

To pick up a flower, participants use the “take” command. For example, Bee picks up flower 2 in the example above. To drop a flower, participants use the “drop” command.

### *5.2 Results*

Table 1 shows that the mean number of moves the participants took to complete the task was 65 and the range was from 44 to 95 moves. Table 1 also shows that some members of a pair made exactly the same number of moves (e.g. pair 1), whilst other members made three times more than their partners (e.g. pair 5).

Figures 2 and 3 show two examples from the study. Figure 2 is the trace of the most successful pair—a pair who solved the problem in 44 moves. The trace diagram clearly shows how they have distributed the task between themselves. One member looked in squares 2–14 and the other looked in squares 11–25.

Figure 3 example is interesting because it is the trace of the least successful pair. They took 95 moves—over twice as many as pair 1 (Figure 1) and 18 moves more than the second least successful pair. Figure 3 shows that this pair visited a lot of squares more than was necessary and often visited squares that their partner had already visited.

### *5.3 Analysis of the Communication*

Table 2 shows that there was quite a range in the amount of paging (i.e. communicating) that participants carried out. Pair 7 only made 1 page between them, compared with pair 9 who made 16 pages. The mean number of pages was 8.

There was no significant relationship between the amount of communication and the number of moves taken to solve the problem ($r = 0.22, n = 9, p > 0.05$). We categorised the communication between the participants into six broad categories: communication about plans, communication about where you are located, commu-

Table 1: Number of moves

| | **Participants** | | |
|---|---|---|---|
| **Pairs** | **A** | **B** | **Total** |
| 1 | 31 | 31 | 62 |
| 2 | 21 | 41 | 62 |
| 3 | 27 | 17 | 44 |
| 4 | 39 | 25 | 64 |
| 5 | 45 | 15 | 60 |
| 6 | 38 | 39 | 77 |
| 7 | 27 | 35 | 62 |
| 8 | 43 | 52 | 95 |
| 9 | 33 | 39 | 72 |
| | | **Mean** | **65** |

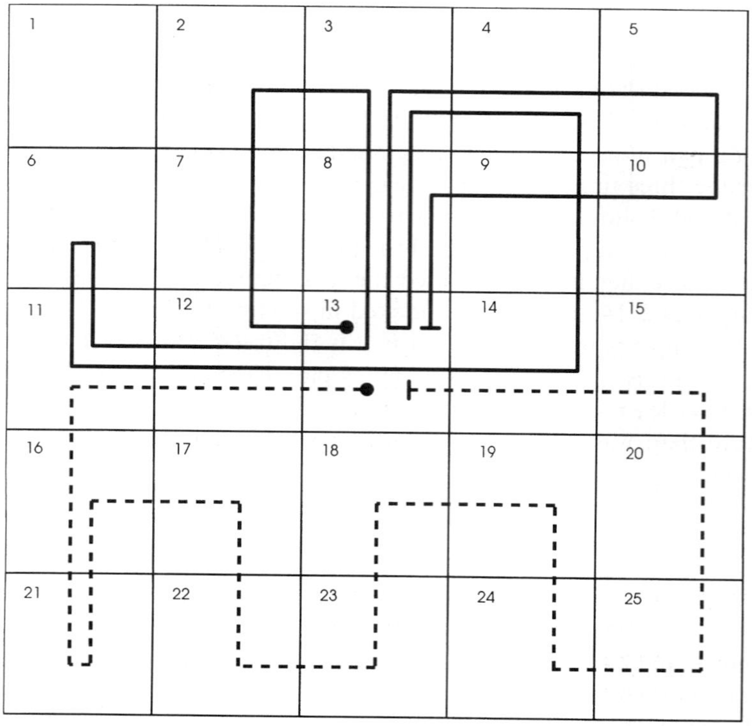


Figure 2: Most successful pair

Table 2: Number of pages

| | Participants | | |
|---|---|---|---|
| Pairs | A | B | Total |
| 1 | 5 | 1 | 6 |
| 2 | 8 | 7 | 15 |
| 3 | 3 | 2 | 5 |
| 4 | 5 | 7 | 12 |
| 5 | 1 | 0 | 1 |
| 6 | 2 | 3 | 5 |
| 7 | 0 | 1 | 1 |
| 8 | 5 | 6 | 11 |
| 9 | 10 | 6 | 16 |
| | | **Mean** | **8** |

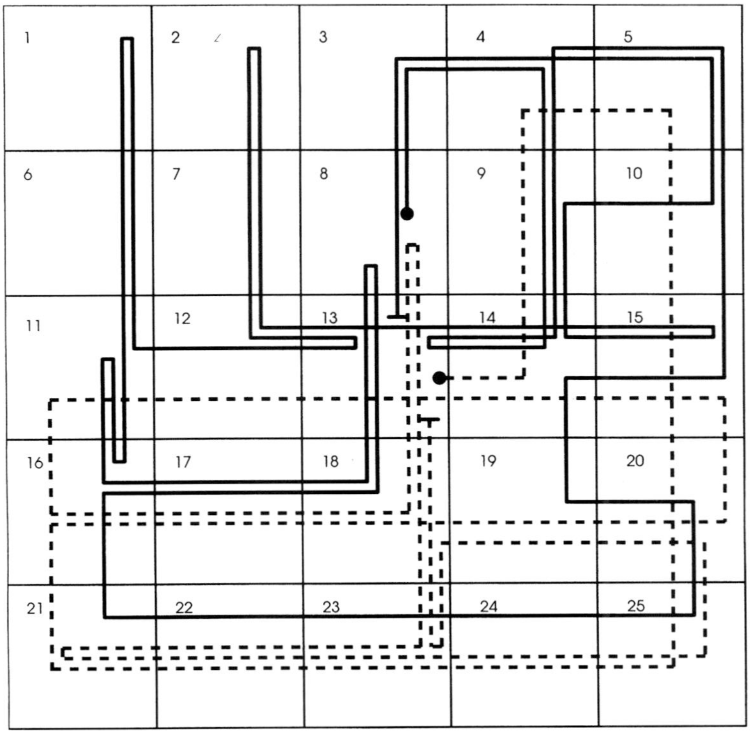


Figure 3: Least successful pair

nication concerning search information, communication about flowers, acknowledgements and miscellaneous. Table 3 shows the number of pages in each category.

### 5.3.1 Plans

Communication about plans were pages concerned with where people were going to search. A third of the pages (i.e. 24) were classified as plans. It mostly occurred at the start of the session. Eight of the nine pairs communicated about how to divide the task up between themselves. Below are three example of how pairs divided the task up.

**Example 1**

Bee You take the upper half and search that, then I'll search the lower half

**Example 2**

Bumble If you go east, I'll go west

Table 3: Classification of pages

| | Talk categories | | | | | | |
|---|---|---|---|---|---|---|---|
| **Pair** | **Plan** | **Location** | **Search** | **Flower** | **Acknowledgements** | **?** | **Total** |
| 1 | 2 | 0 | 0 | 3 | 0 | 1 | 6 |
| 2 | 4 | 1 | 1 | 5 | 3 | 1 | 15 |
| 3 | 2 | 0 | 1 | 1 | 1 | 0 | 5 |
| 4 | 4 | 0 | 3 | 2 | 1 | 2 | 12 |
| 5 | 0 | 0 | 0 | 5 | 0 | 0 | 5 |
| 6 | 1 | 0 | 0 | 0 | 0 | 0 | 1 |
| 7 | 1 | 0 | 0 | 0 | 0 | 0 | 1 |
| 8 | 6 | 0 | 3 | 1 | 0 | 1 | 11 |
| 9 | 4 | 0 | 1 | 7 | 1 | 3 | 16 |
| **Total** | **24** | **1** | **9** | **24** | **6** | **8** | **72** |

**Example 3**

Bee If you search 1-10, I'll do 16-25

Only pair 9 did not communicate about the division of the task. Only pair 8 (least successful pair) made a plan half-way through the session, because they thought they had searched all the squares and still had not found all five flowers.

### 5.3.2 Location

Communication about locations was defined as pages that were explicitly about where the participants were in the world. Only 1 page out of the 72 was classified as communication concerning a person's location. It is shown below.

**Example 1**

Bee I am in room 23

### 5.3.3 Search Information

Search information was communication either about where a participant was planning to search or communication requesting information about where a participant has already searched. Nine pages were classified as being about search information. Three example are shown below. Five of the pairs communicated at some time about search information.

**Example 1**

Bee I've been to room 13-25

**Example 2**

Bumble Have you searched all of the north?
Bee Yes, except 14-15

**Example 3**

Bumble I've looked in 6 and 7

### 5.3.4 Flowers

Communication about flowers was concerned with where flowers were and how many flowers you or your partner had found or had not found. Twenty-four pages were classified as communications regarding flower information. Five pairs communicated at least once about the location of the flowers. Examples 3 and 4 below show the different strategies when pairs realise there are two flowers in room 4: example 3 shows that the initial discoverer decides to pick them both up and example 4 shows that the pair decide to take one each.

**Example 1**

Bumble Flower in room 6

**Example 2**

Bumble Have you found 4 flowers? I've found 1
Bee No, I've only found 3

**Example 3**

Bee There are 2 flowers in room 4, I'll fetch them both

**Example 4**

Bee There are 2 in room 4, come and get one

### 5.3.5 Acknowledgement and Miscellaneous

Six pages were acknowledgements and eight pages did not fit our classification scheme. These were mainly exclamations.

To analyse the relationship between the different communication categories and performance, we correlated the number of moves with the different communication categories. There were no significant correlations between number of moves and communication categories (see Table 4).

Table 4: Relationship between number of moves and communication categories

| Communication | Correlation with no. of moves | Significance ($n = 9$) |
|---|---|---|
| Plan | +0.55 | $p > 0.1$ |
| Location | −0.12 | $p > 0.1$ |
| Search | +0.42 | $p > 0.1$ |
| Flower | −0.10 | $p > 0.1$ |
| Acknowledgements | −0.29 | $p > 0.1$ |
| Unclassified | +0.27 | $p > 0.1$ |
| Total | +0.22 | $p > 0.1$ |

### *5.4 Discussion*

The study we conducted used a task, the methods and means of evaluation normally used in multi-robot research to investigate human–human collaborative search. One of the aims of the study was to investigate human collaborative problem solving with a task that is used with robots. The quickest pair solved it in 44 moves, which is remarkable considering that the shortest number of moves is 38 and that is when you know where the flowers are located. The mean number of moves to solve it was 65, which was mainly due to one pair who solved it in 95 moves. The task is an interesting one to use because it requires the participants to divide the task up between themselves, which is an important collaborative skill, but one that has received very little attention. There have been a number of observational studies (Pozzi, Healey and Hoyles, 1993; Healey, Pozzi & Hoyles, 1995) but no systematic investigations. This type of task provides a means to investigate human's cooperative skills.

The second aim of the study was to evaluate the usefulness of trace diagrams for analysing human–human collaboration—a method of evaluation not used before in studying human collaborative learning. From the three examples of the trace diagrams, it is evident that they provide a useful visual representation of the pairs'' collaborative search strategy. Examples 1 and 2 show clearly that the participants have divided the task up between themselves and example 3 shows that they did not. These trace diagrams could be useful not only in spatial search tasks but also in information search tasks. For example, the joint search of an information database. The database could be represented as a network of locations and the participants search through these locations could be mapped using trace diagrams.

However, trace diagrams do have a number of problems that limit their usefulness for studying human–human collaboration. One is that they do not represent many of the interesting features of human–human collaborations that have been studied: e.g. the nature of the interaction between the participants. It could be the case that a pair simply took turns rather than working collaboratively. Their trace diagram could show little backtracking and overlap, but this would not have been as a result of their collaboration. It is therefore important to supplement this analysis method with other methods to provide a fuller description of their interaction. However, this is true of many paper-based representations of collaborative interactions (Issroff, 1996).

Another problem is that trace diagrams do not accurately represent time. Time is represented by the relative length of the line: the greater the distance along the line, the greater the time from the start. Unfortunately, is not easy to know exactly how much time has elapsed from the start and also where the other participant is at that particular moment. For instance, in example 1, one of the participants stopped searching once he had finished searching his area, and this is not apparent from the trace diagram. A further problem with trace diagrams is that if there were many more moves (i.e. hundreds), the trace diagram would quickly become an unintelligible mass of lines. One possible solution would be to have a number of trace diagrams for different periods of time, but using this method would make it difficult to get an overall picture of the interaction.

The third aim was to investigate the use of communication by humans in this study. In multi-robot systems, Balch and Arkin (1994) showed that performance was improved only if the robots communicated their location to each other and not when they communicated their goals. In this study, only one pair communicated about their location and they only talk about it once. Mostly, the pairs talked about plans, search information and information about the number of flowers they had found or had still to find. From this small study, it is not clear whether this type of communication facilitated performance. The correlation between the amount of talk and performance was very small and not significant. Although it seems essential for the participants to at least agree a collaborative search plan, one pair (pair 5) did not agree a search plan and still managed to solve the problem in the second quickest number of moves. Both of these findings seem to suggest that communication is not that important for this types of task for humans.

Talk in human collaborative learning is thought to be very important and a number of studies have shown this to be the case. The fact that, in this study, it is not clear that it is important is probably due to nature of the task. The two research areas use different tasks because of their differing motivations. Theoretically, certain types of talk are thought to facilitate collaborative learning (i.e. the resolution of conflicts, transactions, co-construction) and it is for these reasons that researchers in human collaborative learning generally pick tasks that are thought to generate a lot of discussion (i.e. physics problems, complex search problems) and do not pick tasks that do not. Multi-robot research picks tasks for engineering or biological reasons, where communication is not that important. In fact, because robots have limited communicative capabilities, there are pragmatic reasons for keeping it to a minimum.

This type of study is potentially useful for multi-robot researchers, as another means for investigating the role of communication in collaborative learning. Multi-robot research could carry out similar studies with different tasks, using humans instead of robots, and if they found, as this study seems to suggest, that communication is not important for intelligent agents with full communicative capabilities, then this would suggest that it is unnecessary to design robots with those capabilities.

## 6 Conclusions

Research concerning collaborative learning between humans and between machines has been gaining a lot of interest in recent years. However, there has been very little contact between the two fields. This paper compared and contrasted the two fields and a study into human's collaborative problem solving was carried out using a task, methods and means of evaluation (trace diagrams) normally employed in distributed artificial intelligence. Pairs of adults had to search a small world for five objects. The results were very interesting in that they showed that trace diagrams could be a useful, albeit limited, tool for studying human collaborative learning. They also seem to suggest that communication was not an important determinant in performance. This chapter illustrates how researchers in both areas can benefit from each other's fields, although care needs to be taken because of the fundamental differences in motivations and the nature of the relevant agents and environments. This is echoed by the chapter title, which refers to the differences between the agents. Human agents design their environments, knowledge, feelings and perceptions. In contrast, robots are designed, and interact in designed environments. This influences the methodologies in both fields.

## Acknowledgements

We would like to thank the European Science Foundation (ESF) Programme on Learning in Humans and Machines for financial support; David Traum for programming the MOO environment; Tia Hansen for arranging and helping with the experiment; and Pierre Dillenbourg and Rolf Ploetzner for constructive criticism of earlier drafts.

# 6

# Learning by Explaining to Oneself and to Others

Rolf Ploetzner, Pierre Dillenbourg, Michael Preier and David Traum

One important source for the acquisition of knowledge, especially of factual knowledge, is the construction, transmission and comprehension of explanations. Two distinctive settings in which explanations are constructed are *self-explanation*, in which a single individual explains to him/herself and *interactive explanation*, in which several individuals mutually and interactively explain to each other. While providing explanations to oneself might lead to the construction of additional knowledge by the explaining individual, providing explanations to each other involves forms of collaborative learning. To understand better the differences between, and potential benefits of, explaining to oneself and explaining to others, we compare empirical research on both kinds of explanation. The comparison is guided by three main goals. The first goal is to describe the learning that takes place during the construction of self-explanations. The second goal is to reveal how the construction of explanations, as well as the learning, changes if one moves from a self-explanation setting to a more interactive explanation setting. With respect to interactive explanation settings, we particularly focus on learning by teaching. The third goal is to uncover differences in the learning between those individuals who construct explanations and those who receive them. Essentially, since learning by teaching has repeatedly been found to promote learning, this gave rise to the assumption that the beneficial effects of such arrangements are due to the successive construction of explanations. In particular, it has been hypothesized that explaining to others might be even more advantageous than explaining to oneself. However, so far, empirical evidence does not strongly support this hypothesis.

## 1 Introduction

When two individuals collaborate, they often have to justify themselves to each other, to explain what they are doing and why they are doing it. Intuitively, these efforts should be related to the learning that is frequently observed during collaboration. For instance, these justifications may lead individuals to make assumptions explicit that would otherwise remain tacit. When an individual explains to a second individual, learning might take place by both individuals:

- The amount of learning by the individual who provides explanations seems to be related to the cognitive activities necessary for constructing and presenting explanations (e.g. Bargh & Schul, 1980; Webb, 1989).

- The amount of learning by the individual who receives explanations seems to be related to variables such as how relevant, understandable and elaborated the explanations are (e.g. Webb, 1989).

In this chapter, we focus on learning due to constructing and providing explanations to oneself or to others. With respect to the settings in which explanations might be constructed, five different levels of interactivity can be distinguished:

1. *Explaining to oneself.* During the attempt to understand something (e.g. instructional material), an individual might try to explain it to him/herself. Explanation is a social criterion for operationalizing what is accepted as understanding. While self-explanations may frequently be expressed silently, only self-explanations expressed aloud can been studied experimentally.
2. *Explaining to a passive and anonymous listener.* An individual might explain to somebody he/she does not know and who just listens. In such a setting, it might be investigated, for example, whether the explainer better monitors the construction of explanations as compared with a self-explanation setting.
3. *Explaining to a passive listener.* An individual might explain to somebody he/she knows and who just listens. In such a setting, it might be examined whether individuals tailor the construction of explanations to specific listeners.
4. *Explaining to somebody who responds in a constrained way.* An individual might explain to somebody who responds to his/her explanations in a constrained way. For instance, the individual who receives the explanations might only indicate his/her understanding or non-understanding. In such a setting, it might be scrutinized how the listener's responses affect the construction of explanations.
5. *Mutual explanation.* Two individuals might mutually explain to each other without any imposed constraints. In this case, explanation is no longer something that is exclusively directed from one individual to a second, but rather corresponds to a process in which two individuals attempt to negotiate and, at least partially, share their understanding of the domain under consideration.

Though only mutual explanation, as described at level 5, accounts for the complexity of many interactions in collaborative problem solving and learning, the investigation of how explanations are constructed at levels 1–4 may nevertheless help to disentangle various processes that might take place almost simultaneously at level 5. Furthermore, fully unconstrained interactions might not necessarily result in the construction of appropriate explanations. Educators, for instance, very often constrain interactions to make them more structured and beneficial (e.g. Aronson, Bridgeman & Geffner, 1978; for an overview, see Knight & Bohlmeyer, 1990).

To understand better the differences between, and potential benefits of, explaining to oneself and explaining to others, we compare empirical research on both kinds of explanation at the different levels of interactivity described above. The comparison is guided by three main goals, as follows. The first goal is to describe the learning that takes place during the construction of self-explanations. The second goal is to reveal how the construction of explanations, as well as the learning, changes if one moves from a self-explanation setting to a more interactive setting. With respect to more interactive explanation settings, we particularly focus on learning by teaching. The third goal is to uncover differences in the learning between those individuals who construct explanations and those who receive them.

Because our comparison is based on a review of various empirical studies and therefore on various perspectives on what makes up explanations, we do not attempt to provide a concise definition of explanatory activities. Instead, we refer to many activities as explanations that in other contexts might be referred to as elaborations or argumentations, for example.

The rest of the chapter is organized as follows. In the next section, we consider self-explanations and describe empirical investigations of their relation to individual problem solving. In addition, we describe *CASCADE*, a cognitive simulation program, which models a possible process by which self-explanations can be constructed during individual problem solving and how they might lead to the acquisition of new knowledge. Next, we consider empirical investigations of explanations in collaborative problem solving and how they might facilitate learning. We conclude with a discussion of (missing) differences in the learning between self-explanation settings and more interactive settings.

## 2 Explaining to Oneself

### *2.1 Revealing Self-Explanations*

How do students learn when they read textbooks, study examples and solve problems? With this question in mind, Chi et al. (1989) investigated nine high school students as they worked on a common college physics textbook (Halliday & Resnick, 1981). Within this investigation, the students carried out the following steps. They read the first three chapters of the textbook, which were on various physics concepts, such as time, distance, velocity and acceleration in the context of motion in one dimension. Afterwards, the students read the textbook's chapter on Newtonian mechanics. The different chapters of the textbook had to be studied until a criterion test had been passed.

After reading the textbook's chapters, the students studied three worked-out examples and solved 25 problems on Newtonian mechanics. A worked-out example consisted of a problem along with a solution that was demonstrated for the students. Figure 1 shows one of the worked-out examples used in the study by Chi et al. (1989). Typically, a worked-out example does not specify all the information that underlies the problem's solution. How does a student know, for instance, that the forces shown in Figure 1 are all the forces acting on the block?

While the students worked on the examples and problems, they were asked to verbalize everything that came to mind. These verbalizations were tape-recorded and later transcribed. Before and after reading the textbook's chapters and working on the examples and problems, the students had to take a pre- and post-test, respectively.

On the basis of the students' scores on problem solving, they were divided *post hoc* into two groups: the four students with the highest scores and the four students with

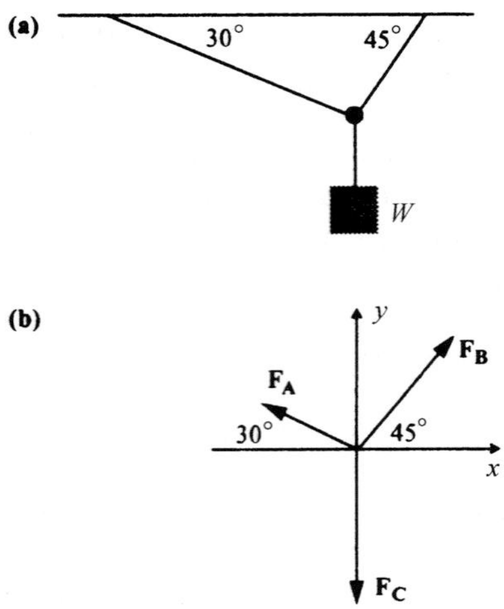


(a) A block of weight $W$ is suspended by strings. Consider the knot at the junction of the three strings to be the body. (b) A free-body diagram shows all the forces acting on the knot. The strings are assumed to be weightless. The body remains at rest under the action of the three forces. Suppose we are given the magnitude of one of these forces. How can we find the magnitudes of the other forces?

$\mathbf{F_A}$, $\mathbf{F_B}$ and $\mathbf{F_C}$ are all the forces acting on the body. Since the body is unaccelerated, $\mathbf{F_B} + \mathbf{F_C} = 0$. Choosing the $x$- and $y$-axes as shown, we can write this vector equation as three scalar equations:

$$F_{Ax} + F_{Bx} = 0$$
$$F_{Ay} + F_{By} + F_{Cy} = 0$$

The third scalar equation for the $z$-axis is simply

$$F_{Az} = F_{Bz} = F_{Cz} = 0$$

That is, the vectors all lie in the $x$–$y$-plane so that they have no $z$-components. From the figure, we see that

$$F_{Ax} = -F_A \cos 30^\circ = -0.866F_A$$
$$F_{Ay} = F_A \sin 30^\circ = 0.500F_A$$
$$F_{Bx} = F_B \cos 45^\circ = 0.707F_B$$
$$F_{By} = F_B \sin 45^\circ = 0.707F_B$$

Also, $F_{Cy} = -F_C = -W$ because the string C merely serves to transmit the force on one end to the junction at its other end. Substituting these results into our original equations, we obtain

$$-0.866F_A + 0.707F_B = 0$$
$$0.500F_A + 0.707F_B - W = 0$$

If we are given the magnitude of any one of these three forces, we can solve these equations for the other two.

Figure 1: A worked-out physics example as used in the study of Chi et al. (1989)

the lowest scores. The data of the student closest to the median were omitted. Since the high- and the low-scoring students performed, on average, equally well on the pre-test, the high-scoring students seemed to have learned more than the low-scoring students. To gain more insight in what distinguishes the high-scoring students from the low-scoring students, the students' verbalizations were classified according to various classes of statements by means of a protocol analysis. The main differences that were revealed by the protocol analysis are:

- The high-scoring students tried more often than the low-scoring students to explain to themselves the different solution steps in the worked-out examples.
- During example studying and problem solving, the high-scoring students were more accurate at assessing their own understanding than the low-scoring students.
- During problem solving, the high-scoring students referred back to the examples less often than the low-scoring subjects. In particular, while the high-scoring students took advantage of only selected aspects of the examples, the low-scoring students frequently re-read the complete examples.

Chi et al. (1989) named the students' attempts to explain to themselves the solution steps in the examples as "self-explanations". Table 1 presents various examples of self-explanations as observed in the study by Chi et al. By means of self-explanations, students refined or generalized the conditions under which solution steps were taken, explicated the meaning of solution steps or extrapolated the consequences of solution steps, for example (see also Chi & VanLehn, 1991). Thus, on the basis of self-explanations, students explain parts of the application domain to themselves and possibly improve their understanding of the domain.

The findings of Chi et al. (1989) and Chi and VanLehn (1991) suggest that self-explanations are constructive cognitive activities that potentially lead to the modification of already available pre-knowledge and the construction of new knowledge. However, could understanding be improved by systematically initiating the construction of self-explanations? Chi et al. (1994) performed an empirical investigation to attempt to answer this question.

Within the investigation of Chi et al. (1994), two groups of students read a slightly shortened section on the human circulatory system from a common biology textbook (Towle, 1989). Before and after reading the textbook section, the students had to take a pre- and post-test, respectively. These tests were made up of four different sets of items:

1. items referring directly to single pieces of information from the textbook,
2. items requiring different pieces of information from the textbook to be integrated,
3. items demanding the induction of new information not provided in the textbook,
4. items requiring the application of information not in the textbook to answer questions about human health.

One group of students, the experimental group, was asked to read each sentence of the textbook section out loud and to explain what it meant to them. The students were instructed to identify new information, to describe how it related to what they already knew and to explain whether it gave them new insights or raised new ques-

Table 1: Examples of self-explanations as observed in the study of Chi et al. (1989); see also Chi and VanLehn (1991)

**Self-explanations related to technical procedures:**

Read (string problem in Figure 1): Choosing the *x*- and *y*-axes as shown, we can write this vector equation as three scalar equations.

Self-explanation: Ummmm, I guess always before when I thought of force vectors I just thought of them as going in a particular direction, I forgot about them having, having *x*-components and *y*-components and being, ummm, broken down into them.

Read (an inclined plane problem): With this choice of coordinates, only one force, *m g*, must be resolved into components in solving the problem.

Self-explanation: I see that because it is the only one that would not be on one of those axes.

**Self-explanations related to physics principles:**

Read (string problem in Figure 1): The body remains at rest under the action of the three forces.

Self-explanation: So the sum of forces should be zero.

Read (an inclined plane problem): Since the block is unaccelerated, we obtain $\mathbf{F} + \mathbf{N} + \mathbf{m}\ \mathbf{g} = 0$.

Self-explanation: There seem to be the three forces involved, also they all sum to zero somehow.

**Self-explanations related to physics concepts:**

Read (string problem in Figure 1): Consider the knot at the junction of the three strings to be the body.

Self-explanation: Uh, so they refer to the point as the body.

Read (string problem in Figure 1): The body remains at rest under the action of the three forces.

Self-explanation: I see. So the *W* will be the force and not the body.

tions. If required, an experimenter reminded the students to explain what they read. The other group of students, the control group, was not asked to explain the textbook section. Instead, in order to keep the time spent on the section approximately constant in both groups, they were asked to read the section twice.

From the pre-test to the post-test, the number of correct answers produced by the experimental group increased significantly more than the number of correct answers produced by the control group. This was especially true with respect to items out of the third and fourth set of test items. These test items required inferences to be drawn and new information to be related to one's own pre-knowledge. It thus appears that understanding can be (partially) improved by systematically initiating the construction of self-explanations.

It might be questioned, however, whether the students' activities in the studies of Chi et al. (1989, 1994) are appropriately denoted as "self-explanations". Essentially, self-explanations as observed in the study of Chi et al. (1989) and as initiated in the study of Chi et al. (1994) are explanations to an experimenter who is (repeatedly) asking for them, though in a constrained way. It remains an open question whether explanations directed (exclusively) to ourselves and explanations directed to others are the same. For instance, are explanations directed to ourselves as complete and consistent as explanations directed to somebody else? Or do we frequently construct only partial self-explanations because nobody else needs to understand them? In order to prepare mentally for important events, such as examinations, it is often helpful to imagine the situation to come and, in particular, to imagine how one would argue in this situation. This suggests that we might adapt our explanations, even when the listener is merely imagined.

### 2.2 *A Cognitive Simulation Model of Constructing Self-Explanations*

To understand better why and how self-explanations lead to the construction of new knowledge, VanLehn, Jones and Chi (1992) developed and implemented the cognitive simulation program CASCADE as a knowledge-based system. In order to formalize the model's initial physics knowledge, they conducted a task analysis. The task analysis comprised two main steps. In the first step, the physics knowledge had been formalized in such a way that it was sufficient for solving all but two of the 25 problems that had been used in the study of Chi et al. (1989). In the second step, it was determined for each piece of physics knowledge whether it could be aligned with information given in the textbook chapters that had been used in the study of Chi et al. (1989). Surprisingly, only about half of the physics knowledge sufficient to solve the problems that had been posed by Chi et al. (1989) was found to be addressed in the textbook chapters. This knowledge made up CASCADE's initial knowledge of physics.

On the basis of its initial physics knowledge, CASCADE models two abilities: the explanation of worked-out examples and the solution of problems. A worked-out example is made up of a problem along with a solution. However, such a solution is hardly ever complete. Usually, only the main solution steps are demonstrated by the example. The explanation of examples, as well as the solution of problems, is modeled in two different ways in order to account for the observed differences between the high-scoring students and the low-scoring students in the study of Chi et al. (1989).

The explanation of worked-out examples by high-scoring students is modeled in the following way. The model attempts to derive each solution step provided within an example by means of the physics knowledge available to the model. Thus, in CASCADE, the construction of an explanation corresponds to the construction of a formal proof. If a derivation can be found, then it is saved for later use during problem solving. If only a partial derivation can be found, then the model tries to complete it by taking advantage of general background knowledge, such as common-sense knowledge, as well as domain-specific and general heuristics. Whenever a derivation can successfully be completed by means of background knowledge, new physics knowledge is constructed that encodes the applied background knowledge in a form constrained according to the information given in the example. Thus, under these conditions, CASCADE realizes a form of explanation-based learning as it has been conceptualized by Mitchell, Keller and Kedar-Cabelli (1986).

If the derivation of a solution step cannot be completed by applying background knowledge, then it is concluded that such a solution step always has to occur within the respective context. Afterwards, new physics knowledge is constructed that encodes the solution step and the context in which it occurred in a generalized form.

The explanation of worked-out examples by low-scoring students is modeled by merely saving each solution step as it is provided in the example. The solution of problems in CASCADE relies on the same mechanisms as the explanation of examples. The model attempts to derive a problem's solution by means of physics knowledge, as well as general background knowledge available to the model. In particular, the construction of new physics knowledge might take place during the explanation of worked-out examples, as well as during the solution of problems.

Problems can also be approached by referring back to the examples. In CASCADE, the use of examples by high-scoring students is modeled by taking advantage of the derivations that have been saved during the explanation of examples. In order to realize a solution step with respect to a posed problem, the physics knowledge that enabled the derivation of an analogous solution step in the example is (re-)applied. In contrast, the use of examples by low-scoring students is modeled by the attempt simply to substitute the information given in the solution steps of the examples for the information given in the posed problems. In many cases, however, the problems posed in the study of Chi et al. (1989) cannot be solved by means of such simple substitution processes.

When CASCADE simulates how high-scoring students explain examples and solve problems, then the model successfully solves all 23 problems that have been taken into account during the task analysis. Furthermore, under this condition, CASCADE learns 23 new pieces of physics knowledge: eight during self-explanation and 15 during problem solving. When CASCADE simulates how low-scoring students explain examples and solve problems, then only nine problems can be solved successfully. Under this condition, the model learns only three new pieces of physics knowledge.

CASCADE simulates how explaining to oneself can lead to the construction of new knowledge. In particular, the model suggests that learning from worked-out examples is most efficient if the solution steps provided within the examples are explained. During the attempt to explain the solution steps, missing knowledge can be identified and subsequently be constructed deductively or inductively by making use of domain-specific and general background knowledge. In many cases, this newly acquired knowledge is a necessary prerequisite for successful problem solving and further learning.

### *2.3 Further Research on Self-Explanations*

The main findings of the study of Chi et al. (1989) have also been found by Ferguson-Hessler and de Jong (1990) in the application domain of electricity and magnetism and by Pirolli and Recker (1994) in the application domain of programming recursive functions in LISP. However, by taking advantage of regression analyses, Pirolli and Recker (1994) demonstrated that the relationship between the amount of constructed explanations and the amount of newly acquired knowledge is not linear. While the construction of self-explanations seems to be related to learning, the learning seems to be more rapid on the basis of the first explanations constructed and less rapid on the basis of the last explanations constructed.

Pirolli and Recker assume that when students work for a long time on a worked-out example, fewer explanations are constructed that lead to the acquisition of new knowledge and more superficial explanations are produced:

> " . . . after the first insightful elaborations are generated, there is a tendency to persevere on paraphrasing or embellishing those early insights without adding anything fundamentally new to the insight and at the expense of explaining other elements." (Pirolli & Recker, 1994, p. 266)

Hence, there appears to be a trade-off where the construction of self-explanations becomes more effective by moving on to the next part of the instruction.

Like Chi et al. (1994), Bielaczyc, Pirolli and Brown (1995) investigated whether understanding in the application domain of LISP-programming can be improved by systematically initiating the construction of self-explanations. Two groups of students were investigated. Both groups worked on the same instructional material. However, after working on an introduction to LISP-programming, one group was trained in constructing self-explanations and making use of self-regulation. This self-regulation training aimed at enabling the students to monitor their problem-solving behavior and to identify and resolve comprehension failures. The other group was not trained to take advantage of these learning strategies. The authors observed that the trained students made significantly more use of self-explanation and self-regulation strategies while working on the instruction than the untrained students. Even more important, this increase in applying learning strategies was also reflected by significantly better programming performances of the trained students.

Renkl (1997a) noticed that in the study of Chi et al. (1994), the group of students who were asked to explain the textbook section spent significantly more time on the instructional material than the group that was not asked to do so. Thus, it could be that the time spent on the instructional material accounts for the observed differences in learning between the two groups. In order to examine this possibility, Renkl (1997a) investigated 36 students as they explained aloud worked-out examples in the application domain of elementary probability theory. Independent of how many examples were studied, the students had 25 minutes time to work on the examples. Before and after studying the examples, the students had to work on a pre- and post-test, respectively.

By means of protocol and regression analyses, Renkl (1997a) found that the amount of constructed explanations correlates positively and significantly with the students' scores on the post-test. This result coheres with the findings of Chi et al.

(1989, 1994). The main result of Renkl (1997a) is, however, that different students predominantly and consistently constructed different kinds of explanations. For instance, while several students predominantly tried to explain the principles that were underlying the examples, other students predominantly tried to explain how the solution steps that were provided in the examples can be achieved. Thus, the concept of self-explanation appears to be less homogeneous than is suggested by the findings of Chi et al. (1989), and it seems that different kinds of self-explanations need to be distinguished.

### *2.4 Summary*

Research on explaining to oneself suggests that self-explanations make up constructive cognitive activities that frequently lead to the acquisition of new knowledge. During the construction of self-explanations, learning seems to take place due to the identification of missing knowledge that would be required in order to complete the self-explanations. Such identified knowledge gaps might subsequently be filled by taking advantage of deductive and/or inductive learning mechanisms. Thus, to request someone to (self-)explain corresponds to asking them to try to understand. Thereby, different kinds of self-explanations might take into account different aspects of the scrutinized application domain and different self-explainers might systematically prefer different kinds of self-explanations.

## 3 Explaining to Others

If explaining to oneself can lead to the acquisition of new knowledge, then explaining to somebody else might have the same beneficial consequences. The idea that the construction of explanations might lead to the acquisition of new knowledge is hardly new; it forms the foundation of the ancient "Socratic teaching" method. In Socratic teaching, the teacher does not teach by direct exposition of the instructional material, but guides the student's own explorations by successively posing questions.

As Webb (1989) points out, explaining to others potentially offers even more opportunities for learning than explaining to oneself. Learning might take place not only due to one's own identification of missing knowledge, but also because the receiver of the explanation identifies further missing information, points out inconsistencies, requires further clarification or confronts the explainer with alternative points of view. In order to resolve these discrepancies, the explainer might search for further information, deduce and induce new information, or restructure already available information, and thus further learn about the domain under consideration. Schwartz (1995) has demonstrated, for instance, that students acquire more abstract knowledge during collaborative problem solving than during individual problem solving. Schwartz assumes that representations constructed during collaborative problem solving frequently need to bridge different viewpoints and thus need to be more abstract than is required for each viewpoint alone.

Explaining to others takes place in almost all situations in which at least two individuals collaborate and communicate. Thus, the construction of explanations is at the heart of collaborative problem solving and learning. However, as Webb (1989) notices, during collaborative problem solving and learning, not all explanations lead to the desired learning effects. Only students who construct elaborated explanations seem to learn from them. This finding is in accord with the findings of research on self-explanation: while the construction of explanations that involve inferences seems to enhance students' knowledge, the construction of explanations that only rephrase what is already known seems to have only minor learning effects.

The main research on explaining to others has been conducted in learning by teaching settings. Within these settings, two or more students or a student and a tutor construct explanations to teach each other.

### *3.1 Reciprocal Teaching*

One important finding of educational as well as psychological research is that students with deficient problem-solving and learning abilities frequently behave rather passively during instruction. If these students could be encouraged to participate actively in instruction, would their understanding improve? With this question in mind, Palincsar and Brown (1984) investigated 24 seventh-grade students as they attempted to acquire various strategies for processing texts, such as locating comprehension failures, questioning, clarifying, summarizing and predicting.

Palincsar and Brown (1984) formed two experimental and two control groups. The two experimental groups received instruction. In order to encourage the students of one experimental group to participate actively in instruction, Palincsar and Brown (1984) developed an instructional procedure called "reciprocal teaching". The second experimental group received instruction by making use of a more traditional method termed "locating information". Both instructional methods are described below. In order to determine the effects of repeated testing, one control group received all of the daily assessments but no instruction. The second control group participated only in the pre- and post-tests.

In reciprocal teaching, an experienced tutor and an inexperienced student take turns discussing a text. The tutor selects a text to be read and indicates whether it is his/her or the student's turn to teach the text. Afterwards, the tutor and the student read the text individually and silently. Thereafter, the teacher (i.e. the tutor or the student) raises questions, points out difficult sections of the text, offers clarifications and explanations, formulates summaries and/or makes predictions about the future content of the text.

Initially, the tutor models the mentioned comprehension activities and the student acts as a rather passive observer. In the beginning, for instance, the student might have severe difficulties when his/her turn to teach comes. However, gradually the student becomes more able to assume his/her role as a teacher and to employ the various comprehension strategies by him/herself. As a consequence, the tutor behaves less as a mentor and more like a partner. Finally, both tutor and student pose questions and offer explanations to each other, as well as engage in negotiations to reach mutual agreement.

According to Palincsar and Brown (1984), the locating information method is frequently used by school teachers to help students in answering questions about a text they have just read. Essentially, the students are shown how to locate information in texts in order to answer specific questions. If information from several text sections needs to be combined, then the teacher demonstrates how to do this. The teacher reinforces correct answers of the students and guides the students back into the text when incorrect answers have been provided. If necessary, the teacher mentions even the lines where the required information can be found.

All students in the study of Palincsar and Brown (1984) had to work on pre- and post-tests that assessed the students' text comprehension. In addition, the students in the two experimental groups, as well as in the first control group, had to work each day on comprehension questions. The students were given feedback on a daily basis. For instance, they were shown diagrams that visualized the percentage of correct answers to the comprehension questions, as well as cumulative records for each week. Overall, the students in the two experimental groups received 20 days of instruction.

The main findings of Palincsar and Brown (1984) can be summarized as follows:

- The students who participated in reciprocal teaching performed much better on the assessments of text comprehension than the students who participated in locating information.
- The students who participated in reciprocal teaching gradually improved in assuming their role as a teacher; their questions and summaries were increasingly expressed in their own words and focused more and more on the main ideas of the text (cf. Table 2).
- The performance on the daily assessments of text comprehension of those students who participated in reciprocal teaching improved in most of the cases from 30% to 80% correctness within 12 days and classroom comprehension scores rose from 20% to 60%.
- The students who participated in locating information performed little better on the assessments of text comprehension than the students who received no instruction at all but all the assessments.

The effects found by Palincsar and Brown (1984) were durable and rather general. Further assessments of text comprehension showed almost no decline in the level of performance for a period of 8 weeks. With respect to the classroom, several students who had participated in reciprocal teaching reached or even surpassed the average level of performance for their age. As a consequence of observing models and teaching, students seem thus to acquire more complete and presumably better-organized knowledge.

Palincsar and Brown (1984) assume that two reasons are responsible for the success of reciprocal teaching. The first reason is that reciprocal teaching involves extensive modeling of the activities to be taught. The second reason is that reciprocal

Table 2: Examples of the acquisition of question asking as observed in the study of Palincsar and Brown (1984) during reciprocal teaching

| | |
|---|---|
| **First day (after reading a short text about snakes):** | |
| Student: | What is found in the southeastern snakes, also the copperhead, rattlesnakes, vipers—they have. I am not doing this right. |
| Teacher: | All right. Do you want to know about the pit vipers? |
| Student: | Yeah. |
| Teacher: | What would be a good question about the pit vipers that starts with the word "why?" |
| Student: | — |
| Teacher: | How about, "Why are the snakes called pit vipers?" |
| Student: | Why do they want to know that they are called pit vipers? |
| Teacher: | Try it again. |
| Student: | Why do they, pit vipers in a pit? |
| Teacher: | How about, "Why do they call the snakes pit vipers?" |
| Student: | Why do they call the snakes pit vipers? |
| Teacher: | There you go! Good for you. |
| **Eleventh day (after reading a short text about the Venus flytrap):** | |
| Student: | What is the most interesting of the insect-eating plants, and where do the plants live at? |
| Teacher: | Two excellent questions! They are both clear and important questions. Ask one at a time now. |

teaching forces the student actively to locate comprehension failures, formulate questions, judge answers to these questions and construct explanations.

### *3.2 Further Research on Learning by Teaching*

In an early investigation, Cloward (1967) had already demonstrated that students can substantially improve their performance in the application domain of text comprehension by means of teaching other students. A group of 240 tenth- and eleventh-grade students with deficiencies in text reading and comprehension tutored a group of fourth- and fifth-grade students who, in comparison with their class mates, also had difficulties in reading and comprehending texts.

Initially, the tutors received training in various teaching activities.[1] Afterwards, each tutor guided one fourth- or fifth-grade student for a period of about 5 months.

1. Unfortunately, Cloward (1967) does not describe which teaching activities have been trained.

In parallel, small groups of tutors met once a week with a supervising teacher in order to discuss problems encountered during the tutorial sessions. A control group of fourth- and fifth-grade students with deficiencies in text reading and comprehension received no tutorial support. Before and after the 5 months of tutoring, respectively, both groups of students had to take a pre- and post-test on text reading and comprehension.

Cloward (1967) found that the ability to read and comprehend texts improved significantly from the pre- to the post-test, not only for the fourth- and fifth-grade students but also for the tutors. Surprisingly, the tutors gained even more than the students who were tutored. Comparable findings are reported by others, including Fantuzzo et al. (1989) and Kafai and Harel (1991a, 1991b); for an overview, see Goodlad and Hirst (1989). In accordance with the results of the study of Palincsar and Brown (1984), these findings also suggest that students acquire more complete and presumably better-organized knowledge by means of teaching. However, what are the reasons for the beneficial effects of teaching? Does the learning mainly take place during preparation to teach, during presentation of the instructional material or during response to questions?

Bargh and Schul (1980) investigated whether students learn while preparing to teach or whether they learn while presenting the material and responding to questions. In a first experiment, after taking a pre-test, one group of undergraduate students studied a text for a fixed period of time to learn it themselves. A second group of undergraduate students studied the same text for the same period of time. Both groups of students were paid according to their scores on a post-test. However, the second group of students were told that, after the study period, they would have to teach the contents of the text to other students. In addition, they were told that the students who received the teaching would take the post-test in their stead and the students' scores would count as their own. Those students who expected to teach gained significantly more than those students who studied only for themselves. Comparable findings are reported by Benware and Deci (1984), for example.

Dayer (1996) observed comparable effects of developing educational software. During the development of courseware, one has to design not only the presentation of information about the application domain but also the presentation of feedback, and thus to anticipate potential errors of those who utilize the courseware. In a study conducted by Dayer (1996), 12 pairs of fourth graders developed simple programs on how plurals of compound names are formed in French. The programs have been developed by means of a graphical authoring tool. Before and after the study, the students had to work on a pre- and post-test, respectively.

Dayer (1996) compared the programs developed by the five pairs with a low gain from the pre- to the post-test with the programs developed by the seven pairs with a high gain. Compared with the pairs with a low gain between pre- and post-test, the pairs with a high gain more frequently included feedback that did not show the correct solution but mentioned the grammatical category of the word to be changed. For example, one pair of students included the feedback "carry is a verb and must hence not be in agreement". Feedback on the grammatical category relates a specific case and a general rule, such as "if the first word is a verb, then it does not take the plural mark".

These results are in accord with Webb's (1989) findings that only elaborated explanations support learning. They might also indicate that the anticipation of the explainee's behavior plays a role equivalent to the reaction to the explainee's actual behavior. However, because the study by Dayer (1996) was of rather preliminary nature, this hypothesis needs to be confirmed in the future by more comprehensive studies.

In a further experiment, Bargh and Schul (1980) scrutinized whether students learn from presenting instructional material and from responding to questions. Three groups of students were investigated. All three groups studied a text as they worked on a problem-solving task for a fixed period of time. One group of students studied alone and silently. The second group of students also studied alone, but were asked to verbalize everything that came to their minds. The third group of students each taught another student while studying. Before and after problem solving and studying, all three groups of students had to take a pre- and post-test, respectively. Bargh and Schul (1980) found no significant differences between the investigated groups with respect to both problem solving and text studying.

Together, these findings suggest that the construction of explanations during teaching, in combination with subsequent feedback, might be less beneficial than one might expect. In order to wade further into the question of whether explanations facilitate learning, Renkl (1995, 1996) investigated two groups of undergraduate students. Initially, both groups of students read an introductory text on elementary probability theory and studied several worked-out examples for a fixed period of time. Afterwards, dyads were formed with one student from each group. One student of the pair was asked to explain several new worked-out examples to the other student for a fixed period of time; the other student was asked to listen. However, the listening student was allowed to pose questions and provide short comments to the explaining student. Before and after studying and then explaining and listening, both groups of students had to take a pre- and post-test, respectively.

In accordance with the results of the study of Bargh and Schul (1980), Renkl (1995, 1996) found that the explaining students did not learn more than the listening students. To further pinpoint how questions affect learning by explaining, Renkl (1997b) investigated two further groups of undergraduate students. Again, both groups of students read an introductory text on elementary probability theory and studied several worked-out examples for a fixed period of time. Afterwards, both groups of students were asked to explain several new worked-out examples to another person for a fixed period of time. However, this time, the person who received the explanations was given specific instruction as to the type of allowable response. In one group, the instructed person posed semi-standardized questions to the explaining students. These questions always had the form "what if . . . ?". In the other group, the instructed person behaved rather passively and provided only neutral acknowledgments such as "hmm" and "I see". Before and after studying as well as explaining, both groups of students had to take a pre- and post-test, respectively.

Renkl (1997b) found that the students who were systematically asked for further information and clarification did not learn more than the students who received simple content-free acknowledgments. Preier (1996) investigated several groups of undergraduate students as they (1) only read a text on the functioning of steam engines, (2) explained the text to themselves, (3) explained the text to other students without receiving feedback from the listeners and (4) explained the text to other students while receiving feedback. Again, no differences were found between students who self-explained and students who explained to others, or between students who explained with or without feedback and students who listened. However, since various aspects were confounded in this study, only preliminary conclusions should be drawn from it.

The categories examined by Preier (1996) are related to the different levels of interactivity as described in the introductory part of this chapter. While no effects were observable in the study by Preier (1996), there is still some reason to believe that different kinds of interactivity can have effects, especially for the receivers of explanations. In a study on task performance rather than learning *per se*, Clark and Schober (1989) demonstrated that "addressees" perform better on a directed task than "overhearers". Using a tangram matching task, pioneered by Krauss and Weinheimer (1964), in which a director explains how to reconstruct a figure before him by using component pieces, two experimental groups tried to perform the task. The first group, the addressees, were allowed to ask questions and interact with the director, giving both positive and negative feedback of their current understanding and task performance. The second group, the overhearers, were allowed to listen to the exchanges of the others, but were not allowed to interact with them. The addressees were not aware that the overhearers were also trying to perform the task.

The results showed that addressees performed significantly better on the task (85–98% correct performance; $F(1, 18) = 10.83, p < 0.01$). Clark and Schober (1989) conclude from these results that the social role of interaction plays a central role in the cognitive process of learning and understanding (see also chapter 3, this volume). Still, it remains to be seen how well this kind of performance-related learning carries over to conceptual learning as investigated in other studies.

### *3.3 Summary*

In research on collaborative problem solving and learning, it has repeatedly been demonstrated that teaching frequently leads to more complete and presumably better-organized knowledge. Since one obvious activity during teaching is explaining to others and responding to questions, this gave rise to the assumption that the beneficial effects of teaching are due to the successive construction of explanations (e.g. Webb, 1989). During the construction of explanations, learning might take place due to one's own identification of missing knowledge, but also because the receiver of the explanation identifies further missing information, points out inconsistencies within the explanation, requires further clarification or confronts the explainer with alternative points of view. However, recently conducted research that aimed at clarifying the role of explanations for learning by teaching does not support this hypothesis. Instead, it has been demonstrated that preparing for teaching, i.e. studying the instructional material individually, results in substantial learning.

## 4 Why Have no Substantial Differences between Explaining to Oneself and Explaining to Others been Observed?

Although, intuitively, explaining to others seems to offer even more opportunities for learning than explaining to oneself, no substantial differences between explaining to others and explaining to oneself have, so far, been observed with respect to learning effects. In the following, we discuss why it might be the case that, until now, no such differences have been observed.

We hypothesize that the empirical research conducted so far did not focus specifically enough on the question of whether explaining to oneself or explaining to others leads to more learning. Especially, it appears that possible differences between the investigated settings have not been maximized, but rather have been blurred. In the research realized so far, various aspects blurred possible differences:

- Self-explanations—as observed, for instance, by Chi et al. (1989)—were not exclusively directed to oneself, but were (mainly?) directed to an experimenter who listened to them.
- Explanations directed to oneself but listened to by an experimenter have possibly been constructed more carefully than explanations directed to peers, because of an experimenter's status.
- In settings for explaining to oneself, students were possibly more systematically prompted to explain than in settings for explaining to others.
- In research such as that conducted by Chi et al. (1989, 1994), students were explicitly asked to construct self-explanations that are beyond simple paraphrases, but relate different kinds of information and give new insight. Conversely, the research on explaining to others did not closely monitor which kinds of explanations were constructed by the students.
- On one hand, it might be the case that successful self-explainers imagined interactions with fictitious explainees and thus imagined explaining and reacting to others. On the other hand, in research on explaining to others, there is often no control over the extent to which an explainer takes into account the reactions of the student who receives the explanations.

It could be that controlling empirical aspects, such as those mentioned above, would allow one to observe more substantial differences between the learning effects of explaining to oneself and explaining to others. However, it could also be that the mechanisms involved in explaining to oneself and explaining to others are not distinct enough or that the effects of these mechanisms are confounded:

- While, during self-explaining, a student might check his own explanations for consistency and completeness; during explaining to others, this task might be shared between the explainer and the explainee. Thus, one might hypothesize that explaining to others results in lower cognitive load and therefore in more learning. However, lower cognitive load does not necessarily lead to more individual learning. Perhaps it is exactly the self-monitoring that plays an important role in learning.
- If explaining to others is to be instructive and efficient, the explainer needs to take into account the explainee's knowledge and understanding (cf. Renkl, 1995, 1996): Does he/she already know what I am going to tell? How shall I explain this? Does he/she have a different point of view? Did he/she understand what I explained? Most probably, such monitoring leads to additional cognitive load on the explainer's side. Again, it is unclear whether this additional cognitive load impedes or improves learning in the long run.
- During explaining to each other, not only does the explainer have a model of the explainee, but the explainee also has a model of the explainer. Messages, including feedback about previous explanations, can affect the models of the receiver, including the receiver's model of the sender's model of the receiver. Such embedded levels of modeling have been studied in linguistic pragmatics, for instance. Clark and Marshall (1981) give examples where deep nestings can be essential in picking out a correct referring expression. Cohen (1978) shows how even infinite nestings can be represented finitely and efficiently. Still, it is not known to what depth such models should be maintained, nor what the precise makeup of such a model is. For instance, Taylor and Carletta (1994) suggest that only two levels of nesting are sufficient for cooperative dialogue. Even if a person models another person sufficiently to continue the dialogue, this might still be merely at a shallow level of understanding, not enough to actually learn. It is possible that such modeling is only taken advantage of to repair communications superficially: e.g. to rephrase an explanation without any further consequences for the knowledge that underlies these explanations.

## 5 Conclusions

Current research does not allow one to take a clear stance on the question of whether explaining to oneself or explaining to others is the more efficient way to learn. However, up to now, almost no research directly addressed this question. More research aimed directly at this question is required to identify the learning effects of both kinds of explaining, and to uncover the processes that lead to this learning. Learning by explaining to oneself and by explaining to others needs to be empirically compared in more systematic ways. In this chapter, we discussed various factors that could be controlled in such comparison studies. Since both explaining to oneself and explaining to others make up constructive cognitive activities, it might turn out, however, that this common characteristic is the one that essentially promotes learning and that it is less important in which of the two settings these activities are realized.

## Acknowledgments

This research has been supported by the European Science Foundation (ESF) within the program "Learning in Humans and Machines".

# 7

# Knowledge Transformations in Agents and Interactions: A Comparison of Machine Learning and Dialogue Operators

Engelbert Mephu Nguifo, Michael Baker and Pierre Dillenbourg

This chapter addresses the problem of understanding the mechanisms by which learning takes place as a result of collaboration between agents. We compare dialogue operators and machine learning operators with a view to understanding how the knowledge that is co-constructed in dialogue can be learned in an individual agent. Machine learning operators make knowledge changes in a knowledge space; dialogue operators are used to represent the way in which knowledge can be co-constructed in dialogue. We describe the degree of overlap between both sets of operators by applying learning operators to an example of dialogue. We review several differences between these two sets of operators: the number of agents, the coverage of strategical aspects and the distance between what one says or hears and what one knows. We discuss the interest of fusing dialogue and learning operators in the case of person–machine cooperative learning and multi-agent learning systems.

## 1 Introduction

What is the specificity of collaborative learning with respect to learning alone? A well-known answer is, for instance, sharing the cognitive load, a mechanism that can be easily translated into a multi-agent architecture. But, the more fundamental point is that, during collaborative learning, different types of interactions occur between learners, some of which have specific cognitive effects. Therefore, a computational model of collaborative learning should integrate the mechanisms that relate interactions with cognitive effects. These mechanisms cannot be a simple "information flow" process, where agent A learns X simply because agent B communicates X to agent A. Such a model would, for instance, contradict the fact that the explainer often gets more benefit than the explainee (chapter 6, this volume). Hence, we are looking for models that deeply integrate dialogue and learning (Dillenbourg, 1996). Specifically, in this chapter, we look for convergences between the basic operators isolated in machine learning (ML) research and in research on dialogue. Beyond this scientific goal, this work also has practical implications: to integrate dialogue operators into ML algorithms to adapt these algorithms, on the one hand, to the interactions with a user (e.g. Mephu Nguifo, 1997), and on the other hand, to interactions with other artificial agents (learning in multi-agent systems).

Thousands of ML systems have been developed, clustered into categories such as "explanation based-learning", "similarity-based learning", "reinforcement learning", etc. Can we describe this variety of algorithms with a restricted set of elementary learning operators? Can we integrate them into multistrategy learning systems? Michalski (1993) addressed these questions. He proposed a set of operators in knowledge space, termed *knowledge transmutations*, which cover the range of inference mechanisms introduced in ML algorithms. Transmutations are generic patterns of knowledge change. A transmutation may change knowledge, derive new knowledge or perform manipulations on knowledge that do not change its content.

This search for "atoms"' within ML is common to many scientific fields, and especially, in dialogue studies. Some researchers attempt to understand the mechanisms by which dialogue leads to learning. They study specific types of interaction associated with learning, such as negotiation (of knowledge, of dialogue focus), argumentation and explanation. Here, we restrict ourselves to considering "knowledge negotiation" or "co-construction of knowledge". In this case, the question arises as to whether we can describe knowledge negotiation in dialogue with an "atomic" set of operators. There exist many classifications of dialogue units, reflecting different theoretical approaches. Here, we use the approach proposed by Baker (1994). This includes a set of operators, termed *transformation functions*, that describe—at the knowledge level—the relations between the contents of utterances produced in collaborative problem-solving dialogues. These operators can be compared with research on content (and other) relations between segments of texts, such as Hobbs" (1982) classification of "coherence relations" in discourse and Mann and Thompson's (1988) "rhetorical relations".[1] There are also many other types of relations that exist between segments of dialogue, other than those that obtain purely on the knowledge level, such as functional relations between speech acts (see, e.g., Levinson, 1983; Moeschler, 1985). What is specific about the approach described in Baker (1994), and particularly adapted to our purposes here, is that it is concerned with relations in dialogue, rather than in text, and that it is based specifically on analysis of collaborative problem-solving interactions.

This chapter compares the two above-mentioned set of operators: Michalski's "transmutations" and Baker's "transformations". They have been chosen as particularly suitable examples, without implying that they are unanimously recognised, in their respective scientific community, as the common reference. We first briefly describe these classifications in, respectively, section 2 for Michalski and section 3 for Baker. We then attempt to apply these operators across disciplines, i.e. to use Michalski's machine learning operators for describing knowledge negotiation in a dialogue (section 4), and to use Baker's dialogic operators on a well-known machine learning problem (section 5). This enables us to explore the extent to which knowledge transformations that take place within a single agent (i.e. machine learning) and during the interaction between two agents can be modelled in terms of similar processes. This leads to a theoretical comparison between the two sets of operators (section 6). Then, in section 7, we draw more practical implications on the interoperability of dialogue and learning operators with respect to the goals stated above:

1. See Sanders, Spooren and Noordman (1992) for a review and synthesis of different categories of textual relations.

modelling collaborative learning and implementing human–machine collaborative learning systems.

## 2 A Taxonomy of Machine Learning Operators

Michalski (1993) defines learning as follows: *given* an input knowledge (I), a goal (G), background knowledge (BK) and a set of transmutations (T), *determine* output knowledge (O) that satisfies the goal, by applying transmutations from the set T to input I and/or background knowledge BK. Transmutations change the knowledge space, i.e. the space where one can represent all possible inputs, all of the learner's background knowledge and all knowledge that the learner can generate. A transmutation may change existing knowledge, derive new knowledge or perform certain manipulations on knowledge that do not change its content.

To define these operators, Michalski introduces two concepts: a reference set and a descriptor. A *reference set* of statements is an entity or a set of entities that these statements describe or refer to. A *descriptor* is an attribute, a relation, or a transformation whose instantiation (value) is used to characterise the reference set or the individual entities in it. For example, consider a statement: "Paul is small, has a Ph.D. in computer science from Montpellier University, and likes skiing". The reference set here is the singleton "Paul". The sentence uses three descriptors: a one-place attribute "height (person)", a binary relation "likes (person, activity)" and a four-place relation "degree-received (person, degree, topic, university)". The reference set and the descriptors are often fixed once and for all in a ML system.

Two categories of transmutations functions are defined:

- *Knowledge generation transmutations* change informational content of the input knowledge. They are performed on statements that have a truth status. These transmutations (see Table 1) are generally based on deductive, inductive, and/or analogical inference.
- *Knowledge manipulation transmutations* are operators that view input knowledge as data or objects to be manipulated. There is no change of the informational content of pieces of knowledge that make up the whole knowledge. Examples are insertion/replication, deletion/destruction, sorting or unsorting operators.

Transmutations are bi-directional operations: they are grouped into pairs of opposite operators, except for derivations that span a range of transmutations.

*Derivations* are knowledge generation transmutations that derive one piece of knowledge from another piece of knowledge (based on some dependency between them), but do not fall into the special categories described above. Because the dependency between knowledge components can range from logical equivalence to random relationship, derivations can be classified on the basis of the strength of dependency into a wide range of forms:

- *Reformulation* transforms a segment of knowledge into a logically equivalent segment of knowledge.

Table 1: Pairs of opposite knowledge generation transmutations (Michalski, 1993)

| | |
|---|---|
| *Generalisation* extends the reference sets of input, i.e. it generates a description that characterises a larger reference set than the input. | *Specialisation* narrows the reference set of objects. |
| *Abstraction* reduces the amount of detail in a description of the given reference set. | *Concretion* generates additional details about the reference set. |
| *Similisation* derives new knowledge about a reference set on the basis of the similarity between this set and another reference set about which the learner has more knowledge. | *Dissimilisation* derives new knowledge on the basis of the lack of similarity between the compared reference sets. |
| *Association* determines a dependency between given entities or descriptions based on the observed facts and/or background knowledge. Dependency may be logical, causal, statistical, temporal, etc. | *Disassociation* asserts a lack of dependency. For example, determining that a given instance is not an example of some concept is a disassociation transmutation. |
| *Selection* is a transmutation that selects a sub-set from a set of entities (a set of knowledge components) that satisfies some criteria: e.g. choosing a sub-set of relevant attributes from a set of candidates, or determining the most plausible hypothesis among a set of candidate hypotheses. | *Generation* generates entities of a given type: e.g. generating an attribute to characterize a given entity, or creating an alternative hypothesis to the one already generated. |
| *Agglomeration* groups entities into larger units according to some goal criterion. If it also hypothesises that the larger units represent general patterns in data, then it is called clustering. | *Decomposition* splits a group (or a structure) of entities into sub-groups according to some goal criterion. |
| *Characterisation* determines a characteristic description of a given set of entities. For example, a simple form of such a description is a list (or a conjunction) of all properties shared by the entities of the given set. | *Discrimination* determines a description that discriminates (distinguishes) the given set of entities from another set of entities. |

- Deductive derivation, abductive explanation and prediction can be viewed as *intermediate derivations*. A weak intermediate derivation is the cross-over operator in a genetic algorithm (Goldberg, 1989). Mathematical or logical transformations of knowledge also represents forms of derivations.
- *Randomisation* transforms one knowledge segment to another one by making random changes: e.g. the mutation operation in a genetic algorithm (Goldberg, 1989).

In the following, we restrict ourselves to the first category (Table 1): changes at the knowledge level, which can later be compared with Baker's knowledge level operators. It is more difficult to relate operators that concern linguistic form since the form of an utterance is very different from the artificial intelligence (AI) knowledge representation scheme.

## 3 A Taxonomy of Dialogue Operators

A key issue in the study of collaborative problem solving is to understand how problem solutions are jointly produced in dialogue. Such common solutions can rarely be reduced to simple "accumulations" of individually proposed solution elements. Rather, solutions *emerge* by an interactive process in which each agent (learner) *transforms* the contributions of the other, in order to attempt to arrive at a mutually satisfactory solution element. This process may be described as one by which *knowledge* is *co-constructed* by a process of *negotiation* (where the term "knowledge" is relativised to the agents concerned, in the absence of a higher authority or arbitrator).

A model for collaborative problem solving in dialogue based on the notion of negotiation has been described by Baker (1994). The model was originally developed for person–machine educational interactions (Baker, 1989); subsequently, it was developed to model collaborative problem solving, having been validated with respect to several dialogue corpora for several different tasks in the domain of physics (Baker, 1995).

Although we cannot discuss this model in detail here, the basic idea is that collaborative problem solving proceeds by a negotiation process, defined as a type of interaction where the agents have the mutual goal of achieving agreement with respect to an as yet unspecified set of *negotia*, under certain constraints (relating to the problem, the social situation, the knowledge states of each agent, etc.). Such a final state may be achieved by three possible strategies: *mutual refinement* (each agent makes proposals, each of which are transformed by the other), *stand pat* (only one agent makes proposals, with different forms of feedback—encouragement, discouragement, etc.—from the other) and *argumentation* (conflict in proposals is made explicit and mutually recognised; each tries to persuade the other to accept their proposals). Although knowledge may, in fact, be more or less indirectly co-constructed during each strategy (e.g. during "constructive argumentation"; see Baker, 1994), here we shall concentrate on the most frequent used and typical strategy: *mutual refinement*.

Each strategy is defined in terms of a set of *communicative acts* and sets of *relations* (created by dialogue operators) that are established between the propositions that they express. The basic communicative acts for the mutual refinement strategy are OFFER and ACCEPTANCE or REJECTION. These are defined using Bunt's (1989) model for dialogue. OFFERs have the following most important pertinence condition (when uttered by agent A1): "accept(A2,p) → accept(A1,p)".

In other words, OFFERs are *conditional* communicative acts that can be interpreted as follows : A1 will accept the proposition p (a problem solution, an action, etc.) *if* A2 will do so ("I will if you will"). Acceptances and rejections have the function of allowing the agent that made the original offer to accept its own offer or not (on the basis that the other does so).

For our purposes here, this view of communicative action in collaborative problem solving has important implications. It means that the information expressed by collaborating students should not be viewed as transmitted knowledge that can be acquired by their partners (*re* the "information flow" view mentioned in the introduction), but rather as the expression of more or less tentative proposals, "to be refined", that will be retained in the common unfolding solution if mutually accepted.

OFFERs and ACCEPTANCEs/REJECTIONs rarely occur in isolation, but rather in sequences, and the sequential position of communicative acts produces additional secondary effects on the contexts of agents. For example, if A1 offers "we are in France", then A2 offers "we are in Lyon", the second offer indirectly communicates acceptance of the first, by virtue of the informational (logico-semantic) *relations* between the contents of the two offers ("Lyon is in France" and "we are in Lyon" → "we are in France"). Similarly, "we are in Lyon", followed by "we are in France" could, in certain contexts, communicate *rejection* (i.e. we are in France, but I do not agree that we are in Lyon), via a Gricean implicature (see later discussion). This is why it is also important to study the relations between communicative acts in this strategy that—at least on the *knowledge* level—may be defined in terms of dialogue operators, or *transformation functions*.

Transformation functions (TFs) are described in terms of the *logico-semantic relations* that are established between the propositions expressed in pairs of communicative acts, either of the same speaker or between speakers. The two communicative acts do not have to directly follow each other in the dialogue. The claim that relations exist between propositions expressed in communicative acts is, of course, a simplification—but one that most often works—since a given proposed proposition relates, in fact, to *the previous context*, from the agent's own point of view. Thus, a given utterance will sometimes relate to what is implicit in a previous one, or what it is taken to imply. This point will be taken up in the discussion.

The following sub-sections summarise the basic TFs, with examples taken from a specific corpus of collaborative problem solving dialogues. The students" task was to find a way of representing the rebound behaviour of balls of different substances (from the experimenter's point of view, the aim was that the students would make some progress towards discovering the coefficient of restitution).

The basic model underlying the sets of TFs is as follows. When two agents, A1 and A2, are engaged in collaborative problem solving, when one agent makes a specific proposal with respect to the problem's solution, the second can respond to it in basically four ways:

1. by *expanding* the proposal (elaborating it further, generalising it, etc.);
2. by *contracting* the proposal (making it more specific, reducing its informational content, etc.);
3. by providing *foundations* for it (justifications, explanations); or
4. by remaining *neutral* with respect to its informational content (either verbatim repetition, or some type of reformulation of the way it is expressed in language, conceptualised, etc.).

### *3.1 Expansion Transformation Functions*

These transform the initially offered proposition by extending or elaborating it in some way, either its "scope" along a generality/specificity dimension, or in informational terms (adding a proposition, inferring a new one on the basis of the initial offer). Table 2 shows typical example TFs in this class.

Often, the transformation takes place in a way that is left partially *implicit*. For example, if one student offers p1, then the second may apply the conjunction TF simply by offering p2; if it is not mutually believed that p2 is contradictory with p1, then the second offer may be interpreted as "OFFER(p2 $\wedge$ p1)".

### *3.2 Contraction Transformation Functions*

These functions are usually the inverse of expansion functions: they restrict the previously offered proposition, or render it more specific (less general). For example, an OFFER "the ball moves at 10 centimetres per second" that follows "the ball has

Table 2: Set of expansion transformation functions (Baker, 1994)

| TF name | Transformation | Examples |
|---|---|---|
| Generalisation | p=TF=> generalisation of p | " . . . the energy . . . " =TF=> " . . . all the energy . . . " |
| Conjunction | p=TF=> p $\wedge$ q | "it's nil on arrival . . . " =TF=> "it's nil at the start, and it's nil on arrival . . . " |
| Specific-value | A(?_x) =TF=> ?_x = $\varphi$ (A—predicate; ?_x—variable) | " . . . kinetic energy" =TF=> "is theoretically nil!" |
| Inference | p=TF=> p $\rightarrow$ q | "potential energy will increase" =TF=> "therefore the rebound force increases" |

slowed down" renders the latter more specific. However, contractions are not always the inverse of expansions. For example, the inverse of inferring a new proposition q from a previous proposition p ($p$ =TF=> $q \rightarrow p$) could be seen as proposing that q implies p ($p$ =TF=> $q \rightarrow p$); in other words, giving a reason or justification for p. This case therefore comes into the *foundational* class (see later). Empirically, the transformation that most often occurs with respect to inferences is to deny their validity. Table 3 shows typical TFs in this class.

The contraction strategy can be viewed as knowledge *deconstruction*. Although, theoretically, inverses of all expansion functions could exist, in reality, examples are hard to find. One possible explanation is in terms of cognitive economy: once the students have co-constructed a possible solution by expansion, if they recognise "that's not it, it can't be that!", it may be easier simply to let the solution drop and start again (perhaps taking part of the previous solution) rather than to deconstruct (contract) it piece-by-piece.

### *3.3 Foundational Transformation Functions*

These provide foundations (reasons for/against, explanations) for offered propositions. Often, this occurs at the end of an expansion phase, when the students "step back" and attempt to verify or check the current joint proposal. For example, in Figure 1:

Table 3: Set of contraction transformation functions (Baker, 1994)

| TF name | Transformation | Example |
|---|---|---|
| Conjunction-contraction | $p \wedge q$ =TF=> $p$ | "the force and the speed . . . " =TF=> " . . . the force . . . " |
| Exclusive-disjunction-choice | $p \vee q$ =TF=> $p$ | "the mass or the density" =TF=> "rather the density" |
| Contra-inference | $p \rightarrow q$ =TF=> $p \wedge \neg q$ | "since it rebounds lower that shows it's the friction" =TF=> "it's lower, but it's not the friction that's involved!" |
| Subtype | $p$ =TF=> subtype $p$ | " . . . all the energy . . . " =TF=> " . . . the kinetic energy . . . " |
| Class-restriction | $p \in \{S\}$ =TF=> $p \in \{\alpha\}, \alpha \supset \{S\}$ | " . . . do you really think a kilogramme ball made of rubber would rebound a lot?" =TF=> "yes, but only in the case of a perfectly elastic rebound" |

| | |
|---|---|
| | (...) |
| [1] A1: | So what can we say if there's an inelastic impact? |
| [2] A2: | Well, that the energy ... all the energy ... |
| [3] A1: | Well, that the kinetic energy is theoretically nil! |
| [4] A2: | It's nil on arrival, in fact ... |
| [5] A1: | Since ... since the object stops, in fact, oh yes, especially since there it doesn't move, uh ... |
| [6] A2: | It's nil at the start and nil on arrival ... about energy ... yes, but at the moment of an inelastic impact, what is it that ... |
| [7] A1: | We've been doing that for a while now! <sighs> |
| [8] A2: | But we've also ... |
| [9] A1: | Wait, ... inelastic impact, right, you've conservation of momentum, but ... the kinetic energy isn't conserved! I think that's what we've seen ... with elastic impact, by contrast, both are conserved |
| [10] A2: | Yes, elastic impact, there's the total energy which is conserved ... |
| [11] A1: | Yes |
| | (...) |

Figure 1: Example of mutual refinement strategy in physics collaborative problem-solving dialogue

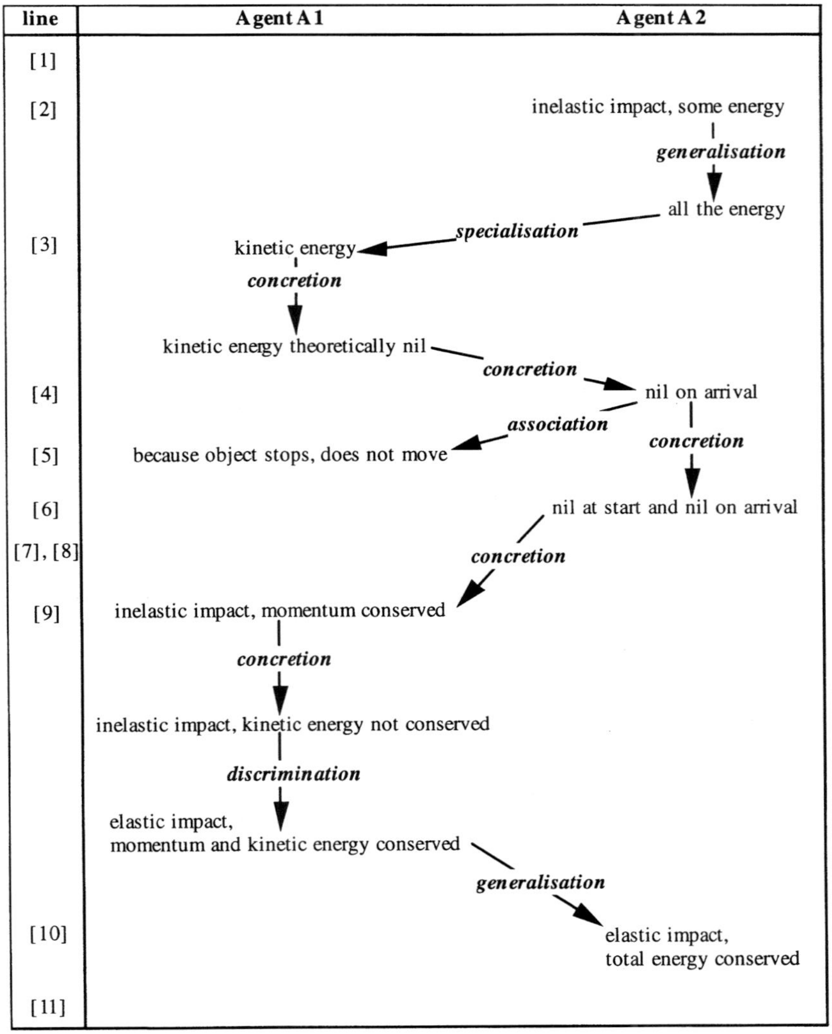


Figure 2: Analysis of example physics collaborative problem solving dialogue using some Michalski's machine leaving operators

"It's nil on arrival, in fact . . . " => " . . . since the object stops . . . "

*Counter-reasons* can indicate a shift to the argumentation strategy, although there may be more isolated occurrences within the mutual refinement strategy when weighing the "pros and cons" of a proposal that is, in fact, mutually agreed.

### 3.4 Neutral Transformation Functions

These either leave the content of the initially offered proposition completely unchanged or else transform its meaning, expression in language or conceptualisation. They often operate at the level of negotiating *understanding* and *agreement, rather than at the knowledge level itself*. This is why exact *repetitions* (nil transformation at knowledge level) usually have the function of confirming understanding and agreement. For example,

" . . . the object stops . . . " = TF=> " . . . it doesn't move . . . "

Often, this occurs as a "termination" to an expansion phase, when the students try to summarise the current joint solution. Some transformations on the level of language (or terminology) are very important from a learning point of view. For example, in

"with the weight" =TF=> " . . . the mass"

the students pass from everyday language and conceptualisations ("weight") to the target (scientific) language/conceptions ("mass").

## 4. Applying Michalski's Machine Learning Operators for Analysing a Dialogue

To evaluate the interoperability of learning and dialogue operators, we attempted to apply Michalski's operators to a short dialogue extract[2] (Figure 1). This illustrates the way in which knowledge is transformed (negotiated) in dialogue using the mutual refinement strategy. Necessarily, the extract illustrates only some of the knowledge transformations or transmutations that could occur. In this short dialogue, the task involves students conducting experiments where they try to discover the properties of balls of different substances that could explain their rebound behaviour—in fact, the coefficient of restitution.

Figure 2 shows a graphical analysis of the extract in Figure 1, using Michalski's knowledge transmutations. The different propositions expressed by each agent (learner) are analysed in separate columns; relations between them, according to the Michalski operators, are shown as labels on arrows.

2. An analysis of the same extract, using transformation functions, is presented in Baker (1994, 1995).

**General Observations.** The extract begins by A1 proposing the focus of the discussion: "inelastic impact" ([1]). This is accepted by A2, and remains the common focus until the second part of [9], when it shifts to the case of "elastic impact". In this first discussion on the case of inelastic impacts, the joint solution is successively transformed, basically by successions of *generalisation* and *specialisation* of what is to be discussed within the "inelastic impact" focus—energy, kinetic energy—and by adding more details to the reference set (class of inelastic impacts) by *concretion*—kinetic energy is nil on arrival at the ground, energy but not momentum is conserved. Once this solution element has been jointly transformed as far as the agents can or want to, A1 then moves on to consider the inverse case—elastic impact.

**Transformations and Collaboration.** The expressed propositions and transformations are distributed across both agents. An important aspect is "who transforms what, expressed by whom?". Thus, agents may transform propositions that they themselves have expressed, either within their turn or across turns. More importantly—for the study of collaborative learning—they transform propositions expressed by others. Such a graphical analysis thus provides one way of mapping out the extent to which agents are really "working together", i.e. collaborating (Baker, 1995). Roschelle and Teasley (1995) have described some similar phenomena as "collaborative produced production rules", distributed across collaborative completions of utterances (the first utterance provides the antecedent of a rule, the second its consequent). When agents are not collaborating (resolving in parallel), this can be seen from the fact that transformations are usually performed by agents on their own contributions across turns.

**What is the Degree of Goodness of Fit?** In each transmutation analysed, the most appropriate operator has been chosen. Nevertheless, certain aspects are left out of the analysis, even when restricting consideration purely to the knowledge level, and more fine-grained distinctions can be made. Thus, within line [5], although agent A1 appears, on the surface, to give a reason for why the kinetic energy is nil on arrival at the ground, this does not really transform knowledge itself into new propositions ("because the object stops, does not move"). Rather, the sequence "nil on arrival"–"object stops"–"does not move" is a sequence of *reformulations* of the meaning of "kinetic energy nil", in more everyday language. A second difficulty with applying the operators can be seen from line [9], where A1 concludes discussion of "inelastic impact" and begins to consider the case of "elastic impact". Is this transformation really discrimination (apparently the best fit), i.e. "determines a description that discriminates the given set of entities from another set of entities"? It is analysed as discrimination here since, by stating that energy is conserved with an elastic impact, this discriminates this case from that of an inelastic impact where energy is *not* conserved. However, in the case considered here, what is important is that the agent moves from considering a class of entities "impacts that are inelastic" to considering the negation of it, i.e. "impacts that are not inelastic (= elastic)". This does not actually *transform* the knowledge base, but rather shifts the focus of attention so that all of the set of "impacts" will be considered. Many of these differences are taken into account by analysis in terms of transformation functions (see Baker, 1995).

**What is Left Out of the Analysis?** In addition to there being finer knowledge-level distinctions that can be made with respect to the propositions that are analysed, some utterances are, of course, left out entirely, since they do not express or transform new knowledge. At this point, it is interesting to mention briefly what they are, for this specific example, and to describe their functions:[3]

*Elements left out of the analysis:*

[1] A1: So what can we say if there's an inelastic impact?

(...)

[7] A1: We've been doing that for a while now!<sighs>

[8] A2: But we've also...

(...)

[10] A2: Yes...

[11] A1: Yes

Line [1] is left out of the analysis since it does not express a new proposition, nor transform knowledge. Rather, it proposes/introduces a new sub-problem to be *focused* on. Lines [7] and [8] express frustration with lack of progress in joint problem solving, from the point of view of A1. In these terms, it can be viewed as part of meta-cognitive control of "knowledge transformation". Finally, lines [10] and [11] act as a "punctuation" to the knowledge transformation sequence. They do this given that the agents must reach agreement, that a sequence of transformations is to be terminated, and that it is so due to joint agreement. All of these aspects missed out of analysis relate to a single fact about dialogue and collaborative problem solving: it needs to be *controlled* or *managed* (Bunt, 1995).

The ML systems use heuristics (background knowledge, types of learned knowledge, generalisation hierarchies, etc.) to control the learning process. These heuristics are generally applied directly to the implemented transmutations functions. This control is purely internal since it appears in a one-shot learning process. As mentioned by Wrobel (1994), these ML systems run like a compiler and do not support the cyclic process of knowledge acquisition or problem solving.

In summary, analysis of collaborative problem-solving dialogues with a specific set of ML operators can, in fact, give us a picture of how collaboration has taken place, on a certain level of generality. Aspects that are specific to (human) interactions, such as negotiation (reformulation) of meaning and interaction management, are, of course, not included in such an analysis. This fact may have important implications given that the effort required to ground interaction, to create joint meaning, may be

3. Note that parts of some utterances that are, in fact, analysed are also left out of consideration here—for example, the "wait . . . " at the beginning of line [9], which relates to coordination of turns. Clearly, we also do not consider purely social aspects of interpersonal interaction, such as maintaining "face", friendship, and so on.

important for collaborative learning (Schwartz, 1995; see also chapter 3, this volume) and that interaction management problems may hinder it.

## 5 Applying Baker's Dialogue Operators to Analysing a Machine Learning Algorithm

In the previous section, we applied machine learning operators to a dialogue. We now perform the inverse, applying the set of dialogue operators to a specific inductive learning algorithm, developed by Mitchell, Utgoff and Banerji (1983). In this case, a concept is built from a set of positive and negative instances. The "version space" is the set of all encountered instances. This space is represented by storing the most specific and the most general definitions of the concept to be learned. Here, we do not enter into the details of the algorithms using this approach, but the principle can be summarised as follows: generalise the most specific description (hereafter S) such that it includes all positive instances, and specialise the most general definition (hereafter G) such that it excludes all negative instances.

### *5.1 Analysis of Example*

In order to articulate this approach with the set of dialogue operators discussed previously, Table 4 presents the case of two artificial agents who conduct an instance-based argumentation (using instances as the main source of evidence) in order to develop their own hypotheses and mutually criticise them. In this fictitious dialogue, agent 1 has a bias towards over-specialisation, maintaining S and criticising G. Conversely, agent 2 is biased towards over-generalisation, maintaining G and criticising S. This example is based on a rudimentary description language—the version spaces S and G are defined by the conjunctive set of (attribute object value) triplets—and a simple generalisation hierarchy. Tautological predicates (e.g. colour object (white or grey)) are removed from descriptions.

Table 4 illustrates the convergence of S and G towards a set of hypotheses (S8 and G7) that match the set of considered instances. Any agent can criticise the hypothesis of its partner (generalising S as in step 2 or specialising G as in step 5), but can also update its own hypothesis as in step 3.

The dialogue is simple because it is assumed that the descrption language and the generalisation hierarchy is shared by the two agents. In a real collaborative learning from examples, those aspects should themselves be grounded in dialogue (see chapter 3, this volume). For instance, the generalisation at step 2 can be either to substitute the class "square" by the larger class "rectangle". An alternative would be to use an exclusive disjunction operator and produce "either a square on the bottom side or a rectangle on the upper side". The dialogue would be richer if the agents had different background knowledge (description language, generalisation hierarchy). Agents should then negotiate between two possible generalisations (e.g. at step 2, between a larger class or the disjunction of smaller classes) or even negotiate a predicate (e.g.

Table 4: Describing an inductive approach to machine learning as a dialogue (~~value or predicate~~ = modified in concept definition, value or predicate = added to concept definition)

| | Instances | Agent | Fictitious dialogue(with dialogue transformation functions) | Evolution of S | Evolution of G |
|---|---|---|---|---|---|
| 1 | Positive 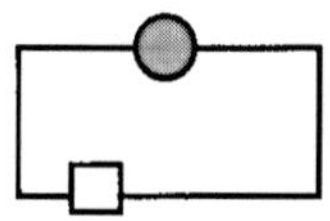 | A1 | *"A Spuc is a rectangle with a small white square on the bottom side and a grey disk on the upper side"* (generates S1) | (shape O1 rectangle)<br>(shape O2 square)<br>(colour O2 white)<br>(size O2) < (size O1)<br>(shape O3 disk)<br>(colour O3 grey)<br>(position O2 (bottom-side O1))<br>(position O3 (up-side O1)) | |
| 2 | Positive 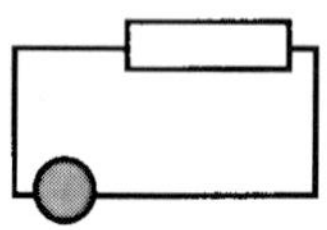 | A2 | *"Actually, your white square can be a rectangle"* (applies generalisation TF) | (shape O1 rectangle)<br>(shape O2 ~~square~~ rectangle)<br>(colour O2 white)<br>(shape O3 disk)<br>(colour O3 grey)<br>(position O2 (bottom-side O1))<br>(position O3 (up-side 01)) | |
| 3 | | A1 | *"OK then but the rectangle can be on any side"* (applies generalisation TF) | (shape O1 rectangle)<br>(shape O2 rectangle)<br>(colour O2 white) (shape O3 disk)<br>(colour O3 grey)<br>(position O2 (bottom-side O1))<br>(position O3 (up-side O1)) | |

| | | | | | |
|---|---|---|---|---|---|
| 4 | | A2 | *"I would just say it is 3 polygons"*<br>(applies generalisation and conjunction-contraction) | | (shape O1 polygon)<br>(shape O2 polygon)<br>(shape O3 polygon) |
| 5 | Negative<br>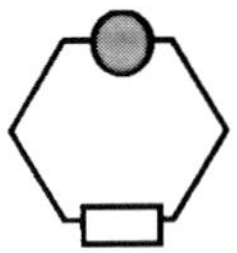 | A1 | *"Not simply 3 polygons: the large polygon must be a rectangle otherwise this instance would be positive"*<br>(applies counter-reason) | | (shape O1 ~~polygon~~ rectangle)<br>(shape O2 polygon)<br>(shape O3 polygon)<br>(size O2/O3) < (size O1) |
| 6 | Positive<br>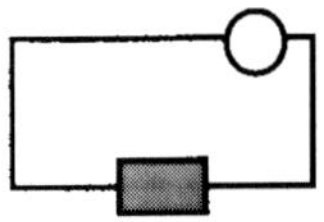 | A2 | "Right but the two small ones can have any colour: here the small rectangle is grey and there the disk is white"<br>(applies generalisation) | (shape O1 rectangle)<br>(shape O2 rectangle)<br>~~(colour O2 white)~~<br>(shape O3 disk)<br>~~(colour O3 grey)~~ | |
| 7 | Négative<br>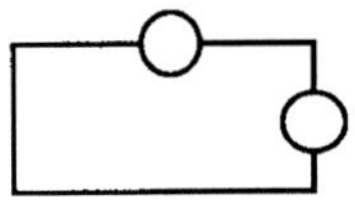 | A1 | *"OK, but you see that the two small polygons must be a disk and a rectangle otherwise this instance would be positive"*<br>(applies sub-type) | | (shape O1 rectangle)<br>(shape O2 ~~polygon~~ rectangle)<br>(shape O3 ~~polygon~~ disk)<br>(size O2) < (size O1) |
| 8 | Negative<br>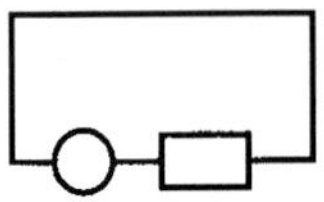 | A1 | *"Moreover they cannot be on the same side otherwise this instance would positive"*<br>(applies specific-value) | (shape O1 rectangle)<br>(shape O2 rectangle)<br>(shape O3 disk)<br>(position O2 (side-X O1))<br>(position O3 (side-Y O1))<br>side-X $\neq$ side-Y | |

agent 1 asks "What do you mean by 'equilibrated'?" and agent 2 answers "Instances 1, 2 and 3 are 'equilibrated' "). The hypotheses S and G would more likely converge if the agents were provided with heuristics that gave priority to attributes that have been negotiated with their partners. Richer interactions would also occur if each agent incompletely ran its algorithm (not exploring all possible generalisations/specialisations), hence generating criticisable hypotheses (e.g. over-general hypothesis in step 4) or providing partial criticisims (as in step 2).

What the example shows is that the overlapping between Michalski's learning operators and Baker's dialogue operators is not suprising given that one can view an inductive approach as the negotiation of the lower and upper limits of the version space.

### 5.2 Comments on the Example

The analysis in section 5.1 shows how simple concept learning can take place through a dialogue between two artificial agents. This occurs through an interplay of three processes: expanding the definition of the concept, contracting it, and invalidating over-generalisations with reference to new instances of the concept.

The process begins with three successive generalisations. In line 2, A2 generalises the description of the square to be "a rectangle"; in line 3, A1 proposes that the rectangle does not necessarily have to be on the upper side, but can be on any side. In fact, these two generalisations of the description are not quite of the same order. In the first case, a single entity (square) is described in terms of the superclass (rectangle); in the second, a determinate description (top side) is universally quantified (the set of all sides in the figure). In line 4, A2 generalises "rectangle" to "polygon", and "contracts" the information about the specific colours of the shapes.

In line 5, A1 invalidates this over-general description, citing the negative instance that shows that the polygon description is too general. In line 6, A2 accepts this invalidation, and goes on to make a specific generalisation about the colour of the smaller objects that had previously corresponded simply to contracting information.

Consideration of the two negative instances in lines 7 and 8 enables the agents to produce a more specific description—the smaller polygons must be a disk and a rectangle (sub-types of class "polygon"), and they must be on different sides.

It is useful to compare this example with the dialogue analysed in Figure 2, where the students were attempting to interpret results of an experiment (two balls of different substances had different rebound behaviours when released above the ground). These two tasks are similar to the exent that the world outside the dialogue provides a set of phenomena that can validate or invalidate their present descriptions of those phenomena/instances. Thus, specific experiments with the rebounding balls correspond to specific presentations of instances. However, the main difference lies in the fact that the students are not told specifically whether the experiments confirm or invalidate their descriptions (interpretations). Often, this is not clear-cut, and is at the origin of much of their dialogue.

## 6 Theoretical Comparison

The analyses presented in the previous sections show that machine learning operators can be applied for describing knowledge refinement in collaborative problem-solving dialogues, and that machine learning can be presented dialogically. However, there are a number of important differences that need to be considered in our comparison. They concern the number of agents, the use of context in communication, the necessity to control dialogue and the relations between agents' mental states and their utterances.

### *6.1 The Number of Agents*

An obvious difference between dialogue operators and ML operators is that the former relate contents of communicative acts uttered by *different* agents, whilst the latter relate knowledge states of the *same* agent. Nevertheless—at the knowledge level at least—this difference is not taken into consideration. First, dialogue operators do function very well as monologue operators, as in line [2] of Figure 2 (A2 generalises his own statement) or in line [3] (A1 concretises his own statement). Conversely, single-agent learning operators can be viewed as dialogue operators in multi-agent learning: an agent A, using operator X to solve a conflict between two divergent goals, can be re-implemented into a multi-agent system, in which two (sub-)agents—A1 and A2—have, respectively, each of these goals and negotiate with the same operator X.

Dialogue and learning operators can be adapted for the case of, respectively, two or one agent, because they are intrinsically binary operators, i.e. operators that describe the differences between two knowledge states (as stored in a system or expressed by agents), without indication as to whether these knowledge entities belong to one or more agents. The distinction between one or more agent is, in any case, not intractable since some distributed cognition theories define a group as a single agent (Salomon, 1993) or, conversely, some scholars view the individual as a group of agents (Minsky, 1987). As pointed by Dillenbourg (chapter 1, this volume), terming something an agent is a theoretical or methodological choice, given that different levels of granularity reveal different aspects of cognitive processes.

### *6.2 Context and Communication*

Whilst in an ML system, the application of a specific transmutation conveys an unambiguous knowledge change; this is not the case when knowledge transformations are expressed as utterances in dialogues between human agents. In fact, in natural dialogues, the apparent, or "surface", knowledge transformation that takes place may not directly correspond to the indirect transformation that is communicated and mutually understood. This is because speakers use the mutually understood context as a means for allowing their hearers to make *inferences*, that will, in turn, correspond to the real knowledge transformation that takes place.

As is well established in pragmatics, the context within which "inferences to meaning" are made must contain general "cooperative principles" (Grice, 1975), as well as knowledge relating to the domain of discourse. Meanings that are conveyed via such principles are termed "implicatures".

Consider the following pair of utterances:

[1] A1: It is a whale
[2] A2: It is a mammal

On the surface level, a *generalisation* transformation has been applied here: the object under discussion, initially described as a whale, is now described as a mammal. There is nothing contradictory between being a mammal and being a whale, since a whale is a mammal. However, in certain dialogue contexts, [2] could, in fact, be interpreted as a *rejection* of [1]. How does this work? Let us assume that A1 and A2 mutually know that a whale is a mammal, and that they both assume the cooperative maxim of "informativeness" ("make your contribution as informative as required"). In this case, [2] appears to be a violation of this maxim, since it states what is already commonly assumed. So, A1 infers that A2 must be intending to communicate something else, i.e. that the object is a mammal, but not (necessarily) a whale. So, the communicated knowledge transformation (invalidation) is not the same as the apparent one (generalisation).

The principle reason that speakers use such indirect means of communication is because this enables greater economy of effort in speech—the context can be assumed and exploited without having to be rendered explicit on every occasion. What does this imply for the possible fusion of dialogue and ML operators? It means that an ML system that incorporates person–machine interaction based on a unique set of operators will be faced with the following choice: either allow implicit language-use for the human user, which would be economical for users but difficult for a machine to understand since it implies maintaining a representation of the context, or else impose a completely explicit dialogue that would be easier for the machine but tiresome for the user. No doubt, an appropriate balance between the two could be found.

### *6.3 The Relation between Knowledge States and Utterances*

We mentioned previously that the knowledge transformation indirectly communicated in dialogue does not necessarily correspond to the transformation that would be proposed from an analysis of the surface contents of successive utterances. A similar problem resides in the fact that the contents of speakers' utterances do not necessarily have direct relations to their mental states.

In general, agents can be viewed as believing what they say, according to a basic assumption of sincerity. However, there is no infallible algorithm for inferring mental states from utterances (although there may be non-monotonic "default" methods; e.g. Perrault, 1990). For example, if one student says " . . . *it's the different densities that explain the rebounds*", what does this mean? That the student *believes* that proposition? Not necessarily. In fact, students often make "proposals"—"might it be x?"—without necessarily having commitment to believing them (Baker, 1994), especially when they are attempting to solve a problem that is beyond their present ability.

Neither do agents necessarily transform the contents of immediately previous utterances. As mentioned above, the agents knowledge spaces are almost never rendered entirely explicit in communication, simply because this would make communication highly uneconomical. Hence, a given utterance often does not relate *directly* to a previous one, but rather to the agent's perception of its underlying context. In the example below, A2 responds, in fact, to the proposition attributed to A1 (what A1 is perceived to be implying/implicating): "the steel ball rebounded higher than the rubber one, so the differences in mass explain the different rebound behaviour".

[1] A1: look, the steel ball rebounded higher than the rubber one
[2] A2: it's not a question of mass!

Finally, agents do not, of course, necessarily believe what they hear. There is a probability that the receiver (1) does not hear correctly, (2) does not understand, (3) understands but disagrees, (4) understands and agrees but cannot integrate in his knowledge base, etc. There is no guarantee that agents directly and completely *internalise* (in a Vygotskian sense; see chapter 2, this volume) the knowledge publicly constructed in dialogue. In this case, if the receiver of an offer accepts it, then the original speaker can infer that both he/she him/herself and the listener can now accept it (see the pertinence condition above). Thus, offers (if accepted) can lead to mutual acceptance.

### *6.4 Dialogue Control and Controlling the Learning Process*

A further gap that appears between dialogue and learning concerns *control*, of both problem solving and dialogue (see section 4). Whilst Michalski's taxonomy does not include "management" operators, control is, however, critical in many learning processes. For instance, in incremental learning from examples, the selection of relevant examples and counter-examples directly influences the learning outcomes. The issue is to know if the operators used in dialogue management and in learning management are similar. The interoperability of learning and dialogue operators should be hence distinctively discussed at two levels: the *knowledge level* and the *control level*.

The existence of implicit meanings and indirect uses of communication indicate that collaborative problem solving in real human dialogues cannot be simply viewed as a direct expression of knowledge transformations. Transformations take place within a dialogue context that is assumed to be mutually understood, but which is rarely completely internalised. Surface transformations in dialogue do not correspond directly to what is *learned*.

These points concern the differences between knowledges states and utterances, rather than the related operators. However, the deep linguistic and philosophical questions raised by the knowledge/utterances discrimination inevitably come into play in the comparison of learning and dialogue operators. However, in the remainder of this chapter, we attempt to see to what extent these theoretical problems can be bypassed for practical goals.

## 7 Practical Implications

### 7.1 Multi-agent Learning

Despite these fundamental issues concerning the gap between mental states and utterances, can the (partial) interoperability of dialogue and learning operators contribute to implement learning in multi-agent systems (chapter 4, this volume)? We examine this issue, with our own perspective where modelling human collaboration prevails on engineering issues.

Unfortunately, dialogue and learning have traditionally been studied in different branches of AI. Traditionally, AI has been clustered into disciplines such as problem solving, ML, computational linguistics, robotics, vision, tutoring systems, etc. Agent architectures reflect this history: they generally dissociate some core reasoning layer (knowledge representation, problem solving, learning, etc.) from the interface layer (incoming and outgoing messages, vision, etc.). There have been a few attempts to merge different techniques and efforts to unify AI sectors, e.g. around SOAR (Rosenbloom, Laird & Newell, 1993), but, globally, the independent development of the different AI disciplines does facilitate synthetic approaches, such as looking for what is common between learning and dialogue models.

The first step to develop an algorithm that unifies learning and dialogue is to improve the interoperability of learning/dialogue operators. People power (Dillenbourg & Self, 1992) illustrates this principle. The artificial agent uses the same operators for reasoning and for dialoguing: agree or refute. Dialogue was based on a binary tree of arguments where any argument could be agreed or refuted, where refutations could then on their turn be agreed or refuted, and so on. When the artificial agent reasoned alone, it used the same operators with itself, i.e. it was able to refute its own arguments, its own refutations, and so forth. In other words, reasoning was implemented as an inner dialogue. Learning resulted from the fact that the artificial agent replayed—*modus modendi*—some parts of previous dialogue during its inner monologues.

In the task, the two agents have to modify an electoral system in order to gain more seats for their party. When "proving" (in the explanation-based learning algorithm (EBL) sense) that his/her action will allow seats to be gained, agent 1 (using over-general rules) argues with agent 2 that: "If a party gets fewer votes, then it will get fewer seats". Agent 2 attempts to prove that this statement is wrong in the particular case at hand and proves it thus: "Since this party has no seats, it cannot lose seats". Agent 1 will store this refutation, and check it the next time it triggers the over-general rule. This relation between the over-general rule and its refutation is functionally equivalent to specialising the over-general rule by adding, as a condition, the negation of the refuting rule: "If a party gets fewer votes and has seats, then it will lose seats".

The principle of similarity between dialogue and reasoning was applied here in its simplest way, the set of operators being extremely simple (agree/disagree). Real dialogues are, of course, more complex. An avenue for research is to design similar dialogue/learning mechanisms but with a richer set of dialogue/learning operators and communicative acts, or systems that are able to negotiate "below the rule level", i.e. not by using conflictual rules, but by negotiating during the rule instantiation process (Dillenbourg, 1996).

The notion of "reasoning as a dialogue with oneself" illustrates the applicability of dialogue operators as learning operators. The reverse applicability, using learning operators to describe group interactions, can be implemented if one sees collaborative dialogues as the process of building a shared knowledge set. Individual utterances can hence be seen as learning operators that transform the shared knowledge set.

However, the shared knowledge space is not a publicly accessible entity, as in the case of knowledge states in an ML system—representations of it exist in each agent. Some discrepancies are actually important for learning, provided that they are detected and resolved. This brings us back to a point mentioned earlier, concerning the multidimensionality of negotiation. In addition to domain-level relations, established by transformations, agents also transform the meaning of utterances, in order to check what was meant, to establish mutual understanding of the joint space—a process that has been described as "social grounding" (Clark & Schaefer, 1989). Such grounding or negotiation of meaning is required even in the case where agents attempted to make all of their relevant knowledge states publicly accessible.

In other words, there is something "stereoscopic" in collaboration: agent A exploits differences between his own knowledge and his representation of agent B's partner knowledge. This interpartner comparison is different from what we have seen so far. Learning operators describe differences between successive knowledge states, dialogue operators between more or less successive utterances. In both cases, this sequentiality limits somewhat the scope of potential differences. Conversely, the differential reasoning on two independent knowledge states must cover any possible difference, until an empty intersection. The set of operators used in both learning and dialogue should hence be extended to cover the mutual modelling process that supports the construction of shared knowledge.

It is likely that the perspective of "multi-agent learning as a dialogue among agents"—which would be a new perspective that has been ignored so far—will open new and promising research directions (chapter 4, this volume).

### *7.2 Integrating User Interaction in Machine Learning*

Machine learning systems most often induce a general concept from a number of examples by generalising the common features of the examples. Among the particular difficulties in such systems are the description language limitations and the incrementality of the learning process. Valid algorithms may produce incorrect results from correct data, simply because there is some mismatch between the way in which data is provided and the way it is processed. This is especially important as we deal with imperfect domain theories, where initial data may be incomplete or may contain many biases. Knowing this, users can modify data that they provide to the system in order to allow it to refine the knowledge that it has already built.

To reduce such difficulties, different ML systems or models have been proposed, and their common purpose is to move from the one-shot learning process, and take into account interaction with expert users. There is a growing interest in integrating interactions with expert users in ML systems. The goal is nevertheless not strictly to implement collaborative learning systems. The main motivation is to gain robustness by adding user–system interactions. Of course, most ML systems allow very restricted types of interactions with the user. These interactions are generally implemented by incorporating the ML system into two kinds of environment:

- *Interface environment.* In such an environment, the systems that have been exhibited tend to place more emphasis on interaction than on learning. These systems belong to the area of "programming by demonstration (PDB)" (Cypher, 1993). The learning systems are used to analyse user behaviour and then try to improve the user's task (see, e.g. Yoshida and Motoda, 1996).
- *Knowledge acquisition environment.* In such an environment, the systems focus more on the learning/acquisition process than on interaction. Knowledge acquisition is viewed as a cyclic process that allows the system to refine its learned or acquired knowledge. Examples of such an environment are the knowledge-based system MOBAL (Morik et al., 1993) and the learning environment LEGAL (Mephu Nguifo, 1997).

LEGAL has its foundations from a learning model reported by Mephu Nguifo (1994, 1995). In such a model, the learning system interacts with an expert user in order to (1) learn knowledge from initial data provided by the expert, (2) derive new results from learned knowledge, and (3) also help the expert user during the interpretation of learning results. LEGAL receives initial input (binary data and learning parameters) from the user. The user is considered to be an expert of the domain. The learned knowledge arises from the successive application of different transmutation functions such as specialisation, selection, characterisation, generalisation and discrimination over the lattice structure of the binary data.

The derivation of new results is a way to validate the learned knowledge. Unseen data are provided by the user to test the efficiency of the learned knowledge. LEGAL uses different kinds of derivation transmutations for this purpose. It can use either deductive derivation or analogical inference by combining reformulation and similisation transmutation functions.

As the user works on an incomplete model, the last step becomes very important since both expert and system can change their knowledge state depending on the results interpretation. To allow this, the learning model includes the notion of proofs and refutations (Lakatos, 1984; Hayes-Roth, 1983) through the mechanism of objections. Objections are built by the system as a way of proving its results. An objection is true until the user refutes it as an explanation of the system decision. Whilst the acceptance of an objection could change the user knowledge, the refutation of an objection should allow modification of the knowledge learned by the system.

This system is based on a static and cyclic process for controlling learning. It has been extended to a dynamic model of control (Mephu Nguifo, 1997). This extended model adds an indirect dialogue between the user and the system (by integrating various tools with different purposes), which are linked together in order to control the learning process dynamically.

Nevertheless, these systems basically remain cooperative (rather than "collaborative", in the sense described here; see also chapter 1, this volume) since there is a fixed division of labour between the system and the user: the former has to learn, explain, or recognise, and the latter must choose the initial data and analyse the system's results. We are seeking more collaborative systems where the user and the system can *negotiate* the data, the amount of noise, the heuristics, etc. In collaboration, the system and the human share roughly the same set of actions (Dillenbourg & Baker, 1996). Hence, any reasoning step performed by the ML system has to be available also to the user through some dialogue function. The interoperability of dialogue and learning operators may improve the collaboration between a human agent and the system agent in two ways: (1) by increasing the modularity of functions, each agent being able to contribute a smaller step of the learning process, (2) by increasing the symmetry of the distribution of functions, most functions being allocable either to the human or to the machine agent. Another potential advantage of incorporating dialogue and learning together is that this might lead to "self-explanation effects" or the "effect of understanding by explaining".

## 8 Conclusions

This chapter compared two sets of operators that come from different research communities. Learning operators have been proposed in ML, where knowledge states are directly inspectable. Dialogue operators come from psychology and linguistics; they describe relations between the content conveyed by the subjects actions—namely, by their utterances—via a commonly assumed context. The gap between dialogue and learning operators reflects the gap between knowledge and action. For instance, we pointed out that Baker's dialogue operators cover the "strategy" aspects of dialogue (e.g. saying something one does not believe in order to check one's partner's agreement), while Michalski's learning operators do not cover these "strategy" aspects (e.g. selecting which example to consider next). While the interoperability of learning and dialogue operators seems feasible at the knowledge level, it seems more difficult to achieve it at the strategical level.

Finally, we mentioned that learning operators could be applied to the analysis of collaborative dialogues, not by modelling individual knowledge states, but by tracing the emergence of a body of shared knowledge. Our recent experiments on computer-supported collaborative work (Dillenbourg & Traum, 1997) show that this shared knowledge is—to a large extent—reified on a whiteboard. A whiteboard is a drawing software in which two users can edit the same page together, modifying or deleting each other's objects. The information displayed on the whiteboard remains there and is mutually visible; the subjects use it as the main medium to co-construct the solution. In other words, they select for display on this persistent medium the information which is itself persistent, i.e. will remain valid along the solution process. Hence, the shared knowledge, this invisible result of grounding dialogues, is reified by a single set of observables notes, which brings this linguistic concept closer to the machine learning framework.

# 8

# Can Analytic Models Support Learning in Groups?

H. Ulrich Hoppe and Rolf Ploetzner

Cognitive models of collaborative learning and problem solving are a major issue of current research in cognitive science. One important question is whether such models can be successfully applied to practical problems of designing collaborative learning environments and of providing intelligent run-time support in group learning situations. After a brief characterisation of the origins and general goals of this research, we delineate a specific knowledge-level model of collaborative problem solving in the application domain of classical mechanics. Afterwards, multiple student modelling is described as a first approach to dynamically supporting problem solving and learning in groups. Finally, the use of analytic cognitive models is discussed in the broader context of designing environments with shared workspaces for collaborative problem solving and learning.

## 1 Introduction: Premises and Promises

Cooperation and collaboration are phenomena for which we still lack adequate analytic models. An analytic model, in this context, is conceived as a *formally represented computational artefact that can be used to simulate, reconstruct, or analyse certain aspects of actions or interactions occurring in groups*. We do not claim and do not even believe that the complex phenomena of social interaction, and particularly of collaborative learning, can be comprehensively simulated or reconstructed by analytic models. Rather, we start from the premise that these models are bound to be partial, in the sense of capturing only specific facets of actions or interactions in groups. We are particularly striving for cognitive models of knowledge representation and knowledge processing. We believe that the development of such models may give rise to interesting new insights and applications, such as the following:

1. the extension of existing cognitive models of individual problem solving and learning to collaborative situations,
2. a better understanding of collaborative problem solving and learning and the identification or specification of relevant determinants,
3. the development of knowledge-level tools for explicitly supporting problem solving and learning in groups, and
4. the specification of design parameters that are to be considered in the construction of computerised environments for collaborative problem solving and learning.

The first three promises can be supported by arguments that are analogous to those used for justifying the codevelopment of cognitive theories and "cognitive architectures" for explaining and simulating individual problem solving, on the one hand, and of intelligent tutoring systems (ITS) that incorporate such architectures to support individual problem solving and learning, on the other hand (cf. Anderson et al., 1990). The last promise is analogous to the role of cognitive theories for *designing* interactive learning environments.

Pursuing this analogy, we could postulate the fusion of CSCL and ITS research into something that could be called *I-CSCL (Intelligent Computer-Supported Collaborative Learning)*. Certainly, research on ITS, and particularly on student modelling, has stimulated the development of cognitive modelling in that the representation of a student model in an ITS has been conceived as a specific cognitive modelling task. In this sense, the more theoretical issue of cognitive modelling has been coupled with the development of student and user models. Accordingly, I-CSCL could play a similar role in fostering the understanding and modelling of problem solving and learning in groups.

However, because ITS have largely failed in making a successful transition from laboratories to educational practice, they are not exactly a success story. And even the very base of cognitive modelling as formulated in the *symbolic information processing hypothesis* (e.g. Simon, 1969) is nowadays being attacked by connectionists and "situationists" (e.g. Clancey, 1992). Indeed, the limitations of symbolic cognitive models have always been acknowledged by a number of researchers (e.g. Ohlsson, 1988). There is no doubt that the shift of focus towards more complex learning situations leaves even more aspects of education, such as social and motivational aspects, uncovered by these models than is the case with individual problem solving and learning. However, there are reasons to assume that this incompleteness might be less problematic for I-CSCL, since ITS are examples of "closed-loop learning environments" in which the system is supposed to embody and control all parameters relevant to the learning process. Practically, ITS are only capable of controlling those parameters that are closely associated with the logic of the application domain. Motivational aspects and other non-domain aspects are usually not treated. In environments that provide intelligent support of problem solving and learning in groups, automatic components can be limited to those aspects in which the computer has its strengths, such as error diagnosis or expert problem solving. Other tasks can be delegated to peers, teachers, or (remote) tutors, for example. In this sense, such environments are "open-loop learning environments" that allow for uncontrolled interactions with the external environment.

The development of innovative CSCL systems and environments is evidently more than a computational problem. On the one hand, CSCL forces us explicitly to reflect social and institutional conditions of learning. In ITS research, these aspects were neglected since the target setting was a supposedly closed world made up of a single learner and a computer running the ITS program. On the other hand, current CSCL research lacks genuine computational approaches: in one extreme, CSCL is conceived as an application of "high-end" multimedia and communication technology. Very often, such technology is made up of tele-learning systems that are straightforward applications of video-conferencing systems with an emphasis on tele-presence. In many cases, however, such systems lack sound pedagogical justifications. In the other extreme, CSCL appears to be a non-technical problem of designing and evaluating social settings and communication structures. Of course, CSCL is frequently conceptualised between these two extremes (e.g. Smith et al., 1991).

The specific challenge of intelligently supported, model-based environments for group learning does not primarily consist in adding some smart components to CSCL, but it lies in developing tools and architectures that reflect the group situation and support the distribution and management of resources between humans and artefacts transparently and intelligently. The analysis of requirements for this complex design task has to draw on the existing body of pedagogical knowledge and experience. It may help us to develop new ways of thinking about roles and interactions in group learning.

In the following sections, we will first identify three dimensions of collaborative problem solving and learning and thus define the target space of cognitive modelling and intelligent support. The cognitive model described in section 3 simulates the complementary use of qualitative and quantitative physics knowledge between two different problem solvers. The next section discusses the possibilities of using analytic models to support group problem solving and learning at run-time and introduces the "multiple student modelling" approach. Yet, intelligent support should only be a second step that complements a carefully designed interactive system. This leads to the issue of integrating intelligent support with group-interactive environments (section 5). Final remarks clarify the potential benefits but also the current limitations of analytically modelling group collaboration.

## 2 Three Dimensions of Collaborative Problem Solving and Learning

Among other dimensions, three cognitive dimensions that appear to be important to the success of collaborative problem solving and learning are (1) the knowledge available to the collaborating students, (2) the information exchanged between the collaborating students and (3) the roles assumed by the collaborating students.

With respect to the first dimension, the knowledge available to the collaborating students might range from completely disjointed and rather unrelated knowledge, to disjointed but complementary knowledge, to partially overlapping or (almost) identical knowledge. For instance, because mutual interdependence among the collaborating students seems to positively influence collaborative problem solving and learning, different students are sometimes intentionally taught different aspects of the application domain under consideration before collaborative problem solving takes place (e.g. Aronson et al., 1978; Ploetzner et al. 1996a).

With respect to the second dimension, the information exchanged between the collaborating students may range from the mere exchange of achieved results, over the exchange of single specific or general pieces of knowledge, to the exchange of complete specific or general explanation structures. While learning from the exchange of achieved results might not be very far-reaching (e.g. Webb, 1989), the exchange of complete and highly elaborated explanation structures might easily overburden the memory and processing capacities of one's partner.

With respect to the third dimension, students may assume fixed roles during collaborative problem solving and learning, may switch roles on various occasions, may not assume any roles but nevertheless collaborate with constraints imposed on their interaction or may collaborate without any constraints imposed on their interaction. For instance, while one student might assume the role of somebody who teaches and explains, a second student might assume the role of somebody who learns and raises questions (e.g. Palincsar & Brown, 1984; see also chapter 6, this volume).

Together, the three dimensions described make up a space of possible arrangements for collaborative problem solving and learning. We aim at developing a cognitive simulation model of collaborative problem solving and learning that can be set up according to the different points in the space of possible arrangements. The first goal of our research is to obtain deeper insight into which points in the space of possible arrangements make collaborative problem solving and learning both instructive and efficient and which do not. The second goal of our research is to explore how a cognitive simulation model could be taken advantage of in a computer-supported environment for collaborative problem solving and learning.

Within such an environment, a cognitive model of collaborative problem solving and learning could possibly be brought into play in the following ways:

- Given a diagnosis of the knowledge available to the collaborating students, the simulation can be used to select problems that require the collaboration of two or more students.
- Given a diagnosis of the knowledge available to the collaborating students and the roles assigned to them, the simulation can be utilised to select problems that can—at least, in principle—be successfully approached by the students.
- Given a problem to be solved collaboratively, the simulation can be taken advantage of to monitor when turn-taking and role-switching among the collaborating students should ideally take place.

## 3 A Cognitive Simulation Model of Collaborative Problem Solving

Two factors that seem to positively influence collaborative problem solving and learning are (1) task structures that promote mutual interdependence among the collaborating students (e.g. Knight & Bohlmeyer, 1990) and (2) communication structures that promote the construction of high-level questions and explanations (e.g. Webb, 1989). As a first step towards a cognitive model that simulates collaborative problem solving and learning along the three dimensions described above, Ploetzner et al. (1997) developed a cognitive simulation model of collaborative problem solving that allows one to take both factors into account.

The model rests on two basic assumptions. The first assumption is that collaborative problem solving is especially advantageous to the involved subjects if they initially have different perspectives on the considered application domain (e.g. Doise &

Mugny, 1984; Howe, Tolmie & Rodgers, 1990, 1992). The second assumption is that collaborative problem solving requires the ability to carry out adequate self-diagnoses in order to be able to make appropriate use of one's partner as an information resource. The model comprises two simulated problem solvers (see Figure 1). Each problem solver is made up of a cognitive level and a meta-cognitive level.

The cognitive level was realised by taking advantage of a knowledge-based system of individual problem solving in classical mechanics (cf. Ploetzner, 1995). The application domain is made up of problems that involve one-dimensional motion with constant acceleration. The focus is on problems that ask for a precise quantitative problem solution. For example, a block of mass $m = 15$ kilograms starts from rest (initial velocity $v = 0$ meters/second) down a frictionless plane (coefficient of friction $f = 0$) inclined at an angle $\alpha = 30°$. What is the block's velocity $v$ after time $t = 2$ seconds?

The two simulated problem solvers differ only with respect to the knowledge they embody on the cognitive level. While one problem solver embodies exclusively knowledge about qualitative aspects of classical mechanics, the other problem solver embodies exclusively knowledge about quantitative aspects. Both problem solvers make use of the same mechanisms to interpret the knowledge available to them.

Qualitative aspects of physics concepts are formalised by means of two kinds of expressions. The first states the conditions under which a physics concept can legitimately be applied. The second constrains the values of the various attributes that a physics concept has. Knowledge about quantitative aspects of physics concepts is formalised by means of expressions that describe algebraic equations.

Table 1 shows how knowledge about qualitative and quantitative aspects of the normal force is formalised by means of an equation-based representation language similar to the representation language employed by VanLehn, Jones and Chi (1992). Time-dependency of physics parameters is represented by means of an extensional

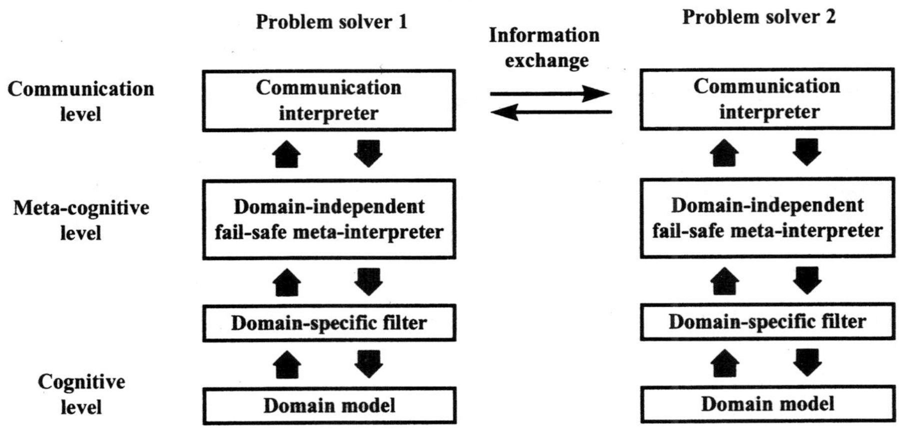


Figure 1: The architecture of the cognitive simulation model

Table 1: How knowledge about qualitative and quantitative aspects of the normal force has been formalised

**Knowledge about qualitative aspects:**

$F_{n1}$: if
value(S, instance($Object_1$, body)) = true and
value(S, instance(force($Object_1$, earth, fg), gravitational_force)) = true and
value(S, instance($Object_2$, surface)) = true and
value(S, on($Object_1$, $Object_2$)) = true
then
value(S, instance(force($Object_1$, $Object_2$, fn), normal_force)) = true

$F_{n2}$: if
value(S, instance(force($Object_1$, $Object_2$, fn), normal_force)) = true and
value(S, instance($Object_2$, surface)) = true
then
value(S, inclination(force($Object_1$, $Object_2$, fn))) =
perpendicular(value(S, inclination($Object_2$)))

$F_{n3}$: if
value(S, instance(force($Object_1$, $Object_2$, fn), normal_force)) = true and
value(S, instance($Object_2$, surface)) = true
then
value(S, sense(force($Object_1$, $Object_2$, fn)) = value(S, relative_position($Object_1$, $Object_2$))

**Knowledge about quantitative aspects:**

$F_{n4}$: value(S, magnitude(force($Object_1$, $Object_2$, fn))) =
value(S, magnitude(force($Object_1$, earth, fg))) * cos(value(S, inclination ($Object_2$)))

temporal operator "value(S, P)", which denotes the value of a parameter "P" in situation "S". The situation might refer to a point in time or an interval of time.

The cognitive, meta-cognitive and communication levels of the two simulated problem solvers were implemented in PROLOG. In PROLOG, problem solving corresponds to the attempt to construct a proof tree by means of the knowledge available on the cognitive level. If the knowledge available to a problem solver is incomplete with respect to a given problem, then no complete proof tree can be constructed and problem solving fails. This is exactly what happens if one of the two simulated problem solvers is applied to a problem whose solution demands the coordination of knowledge about both qualitative and quantitative aspects. The main purpose of complementing the cognitive level with a meta-cognitive level was to provide a domain-independent means that enables (self-)diagnoses of (1) where an impasse is reached during the attempt to construct a proof tree and (2) which information would allow one to overcome such an impasse and possibly to resume the construction of the proof tree. Obviously, if such (self-)diagnoses could be conducted, they would provide an informative basis for posing questions to one's partner.

The meta-cognitive level was implemented by taking advantage of meta-programming techniques that have been developed in the framework of student modelling (cf. Beller & Hoppe, 1993; Hoppe, 1994). Its implementation started from a standard meta-interpreter for PROLOG programs (e.g. Sterling & Shapiro, 1994). In the next step, this interpreter has been made "fail-safe". The fail-safe meta-interpreter marks unprovable goals as failed and includes them into the proof tree without terminating the deduction process. It is complete with respect to the construction of complete proof trees: if it is possible to construct complete proof trees, then all of them are constructed before any incomplete proof tree is constructed.

However, the fail-safe meta-interpreter is not complete with respect to the construction of incomplete proof trees. There might exist incomplete proof trees that cannot be constructed by the meta-interpreter. If no complete proof tree can be constructed, then, in general, the notion of a goal that fails is ambiguous: if a certain goal on the cognitive level fails, then every goal in the (incomplete) proof tree that is hierarchically above the failing goal fails too. However, there exist always lowest level goals within the hierarchical sequence of failed goals. These lowest level goals allow for generating maximally specific questions.

If a simulated problem solver cannot solve a given problem on the cognitive level, the meta-interpreter passes an incomplete proof tree to the communication interpreter. The communication interpreter filters out those goals that failed and forwards them as questions to the other problem solver via the standard TCP/IP communication protocol. The problem solver that receives questions treats them as new problems and attempts to solve them. In its current implementation, the communication interpreter allows only for non-overlapping problem solving and asymmetric communication. Non-overlapping problem solving refers to the constraint that only one problem solver can be active at a time. Asymmetric communication denotes the constraint that a problem solver that attempts to answer a question cannot, in turn, ask the other problem solver for further information. In the future, the model will be equipped with a more flexible communication protocol and manager.

Table 2 shows the performance of the two simulated problem solvers while attempting collaboratively to solve the problem described above. The different problem-solving steps shown in Table 2 are traced on the cognitive level. The different questions and answers shown in Table 2 are traced on the communication level. In the following, the problem solver that comprises exclusively knowledge about qualitative aspects is named the "qualitative solver". The problem solver that comprises exclusively knowledge about quantitative aspects is named the "quantitative solver".

By convention, after both solvers are initialised with the given and sought quantities of the problem, the quantitative solver attempts to solve the problem. Essentially, on the cognitive level, problem solving relies on backward chaining. In a first step, a kinematic equation is considered to determine the block's velocity (cf. expression 1 in Table 2). Because acceleration is a further unknown quantity in this equation, Newton's second law is considered next to determine the acceleration (cf. expression 2). However, because the quantitative solver has no knowledge about qualitative aspects, it cannot determine the resultant force on the block. Therefore, a different kinematic equation is tried out to determine the block's velocity (cf.

Table 2: The performance of the qualitative (white areas) and quantitative (grey-shaded areas) solver

| | | |
|---|---|---|
| | trying equation v = v0 + a*t for value(2,magnitude(velocity(block))) | 1 |
| | trying equation fr = m*a for value(1,magnitude(acceleration(block))) | 2 |
| | trying equation exp(v,2) = exp(v0,2) + 2*a*r for value(2,magnitude(velocity(block))) | 3 |
| | trying equation fr = m*a for value(1,magnitude(acceleration(block))) | 4 |
| **Query:** | value (1,magnitude(force(block,resultant_force,fr)))? | 5 |
| | derived value(1,instance(mass(block),mass)) = true | 6 |
| | derived value(1,instance(force(block,earth,fg),grav_force)) = true | 7 |
| | derived value(1,instance(force(block,plane,fn),normal_force)) = true | 8 |
| | . . . | |
| | derived value(1,arrows(block)) = [arrow(force(block,earth,fg),90,down), arrow(force(block,plane,fn),120,up),arrow(velocity(block),30,down)] | 9 |
| | derived value(1,instance(coordinate_system(block,[axis(x,30),axis(y,120)]),coord_sys)) = true | 10 |
| | derived value(1,instance(free_body_diagram(coordinate_system(block,[axis(x,30),]), axis(y,120)]),[arrow(force(block,earth,fg),90,down),arrow(force(block,plane,fn),120,up), arrow(velocity(block),30,down)]),fbd)) = true | 11 |
| | derived value(1,magnitude(projection(acceleration(block),axis(x,30))))*value(1,mass(block)) = value(1,magnitude(projection(force(block,earth,fg),axis(x,30)))) | 12 |
| **Answer:** | value(1,magnitude(force(block,resultant_force,fr))) = value(1,magnitude(projection(force(block, earth,fg),axis(x,30))))! | 13 |

```
          . . .
Query:    value(1,instance(_20,free_body_diagram))?                                                   14
Answer:   value(1,instance(free_body_diagram(coordinate_system(block,[axis(x,30),axis(y,120)]),       15
             [arrow(force(block,earth,fg),90,dow),arrow(force(block,plane,fn),120,up),
             arrow(velocity (block),30,down)]),fbd)),[],true)!
          . . .
          trying equation fr=m*a for value(1,magnitude(acceleration(block)))                          16
Query:    value(1,angle_between(axis(x,30),arrow(force(block,earth,fg),90,down)))?                    17
          value(1,angle_between(axis(x,30),arrow(force(block,earth,fg),90,down)))=60!                 18
Answer:
          . . .
          derived value(1,magnitude(force(block,earth,fg)))=15*c(g)                                   19
          derived value(1,magnitude(projection(force(block,earth,fg),axis(x,30))))=15*c(g)*cos(60)    20
          derived value(1,magnitude(acceleration(block)))=cos(60)*c(g)                                21
          derived value(2,magnitude(velocity(block)))=cos(60)*c(g)*2                                  22
Solution: value(2,magnitude(velocity(block)))=cos(60)*c(g)*2                                          23
```

expression 3). This leads, again, to an impasse when the resultant force on the block has to be determined (cf. expression 4). Because no further kinematic equations are available, the quantitative solver cannot determine the block's velocity on its own.

In this case, the "lowest" goal that fails on the cognitive level of the quantitative solver is a goal that aims at determining the resultant force on the block. Accordingly, the quantitative solver asks the qualitative solver for the resultant force (cf. expression 5). In order to determine the resultant force on the block, the qualitative solver constructs a so-called (symbolic) free-body diagram: it establishes all forces on the block (cf. expressions 6–8), represents the forces—as well as the block's velocity—by means of arrows (cf. expression 9) and establishes a coordinate system made up of two coordinate axes (cf. expressions 10 and 11). Thereafter, the forces are projected onto the axes. Finally, it is concluded that the resultant force on the block equals the projection of the gravitational force onto the $x$-axis (cf. expressions 12 and 13).

After the quantitative solver received the answer, it attempts to determine the magnitude of the resultant force on the block. However, the quantitative solver fails, because it lacks further information about the free-body diagram and its axes. Therefore, the quantitative solver asks for the free-body diagram (cf. expression 14). The qualitative solver can answer this question directly, because the free-body diagram has already been constructed (cf. expression 15). Having received the free-body diagram, the quantitative solver can still not apply its trigonometric laws to determine the magnitude of the resultant force (cf. expression 16). This failure occurs because the quantitative solver does not know where the angle of inclination is located in the free-body diagram. By convention, this knowledge was made available to the qualitative solver. Thus, a last question is posed to the qualitative solver (cf. expression 17). Finally, after having received the answer, the quantitative solver straightforwardly determines the problem's solution.

The model described above provides insight into some of the processes that may underlie reasoning and exchange processes in collaborative situations. Such a model can be utilised to support the design of group arrangements, learning materials, and problems. In computational terms, such a model supports "static" design decisions as opposed to dynamic reactions to certain specific situations. In the following, we turn our attention to the possibility of providing real-time, on-line support in collaborative problem solving and learning based on analytic modelling techniques.

## 4 Model-Based Support for Group Learning

The on-line analysis of group interactions aiming at assessing and updating individual learner models of the group members appears to be currently intractable for unrestricted interaction, including (spoken) natural language or even non-verbal communication. Even if the group interaction leads to concrete actions and state changes in shared workspace environments, it is often not evident how to assign individual credits for concrete achievements or failures (cf. chapter 5, this volume). Following a recent suggestion by Paiva (1997), this difficulty can be avoided by treating the entire group as a subject of the modelling process according to the principle that "the group is more than the sum of its parts". In Paiva's (1997) approach, the group model should be built upon actions and beliefs on which the group members have agreed. However, with unrestricted interaction, the tractability problem is now shifted towards defining operational criteria for agreement. Also, it is not clear how to initialise the group model when the collaborative learning session starts after a phase of individual work.

We believe that the general problem of analysing rich group interactions will be an issue of research for years to come, and it may even turn out that certain aspects of the problem are indeed intractable. To avoid the complexity of analysing realistic group interactions, we suggest starting by using individually assessed student models to "parameterise" group learning situations (Hoppe, 1995). To make this more modest approach widely applicable, it is desirable that the extrapolation of individual student models not be dependent on a special modelling technique. So, we depart from general information that is essentially provided by any type of student model. Accordingly, we do not make specific assumptions about the actual assessment of the model, i.e. about the diagnostic procedure.

Individual tutoring ideally takes place in a "closed loop", in which every utterance or action of the tutee is interpreted and can lead, if considered useful by the system, to specific feedback. We have already argued that, practically, this is not feasible in rich group interactions. Thus, in intelligently supported group learning we do not expect to have a closed feedback loop, but an open system that evaluates and intervenes only under certain specific conditions. This general framework is visualised in Figure 2.

Open-loop learning with a learning support system is indeed not only specific to group situations. It is also typical for free individual exploration in interactive learning environments (ILE) or micro-worlds. Here, the role of the environments consists in providing a rich "activity space" that, implicitly rather than explicitly, creates opportunities for learning. However, if given a limited, non-comprehensive function in evaluating and supporting the learning process, student modelling can also play a

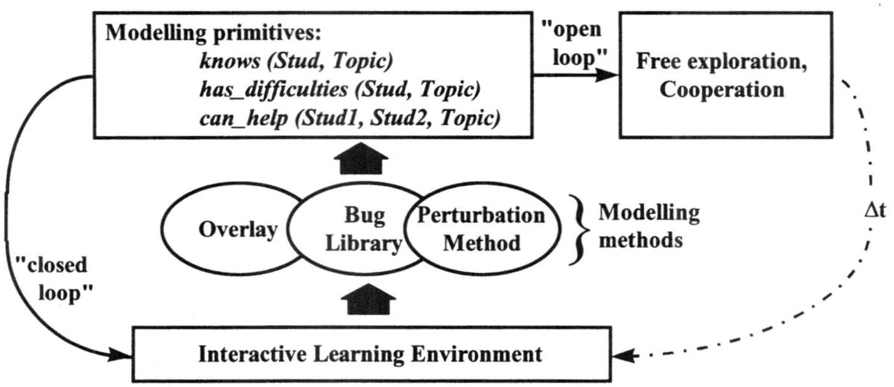


Figure 2: Open framework for student modelling (Hoppe, 1995)

role in ILE (Self, 1994). Phases of more or less free or unguided exploration may, for example, be followed by exercises with closed-loop feedback.

As for supporting group learning, parameterisation of group interaction based on individual models is a tractable instantiation of this open framework: the on-line modelling system does only evaluate individual performance, for instance, in phases of individual routine exercises. The results are used to "inform" group learning scenarios, such as *tutorial hotlines*, *peer help* or the "*ad hoc*" formation of working groups. In these scenarios, intelligent support will mainly be invoked explicitly on demand by one or several of the participants. Though learner modelling for groups of learners (and possibly a human tutor or teacher) poses new problems, the availability of human support and peer-to-peer cooperation can also help to avoid existing problems in generating adequate, personally meaningful feedback to learners. Generally speaking, intelligent sub-systems may help in the tasks of knowledge assessment and error diagnosis, whereas the actual tutoring may be left to human–human interaction of different types. Even in the case of a misdiagnosis, we can expect human–human communication to provide more flexible and ultimately more successful repair strategies (Suchman, 1987).

There is a wide range of different support functions that can be implemented in this modelling framework:

- *Intelligently mediated peer help*. Here, the individually assessed learner models are used to match pairs of learners that should maximally benefit from each other when working together. The prediction can be based on different criteria, such as complementarity of skills and knowledge or competition. A first implementation of this principle has been reported in Hoppe (1995). Practical applications are now being developed (e.g. McCalla et al., 1997).
- *Intelligently mediated expert tutoring*. Formally, this case can be considered as a specialisation and simplification of matching peer learners, since only one of the models—namely, the learner's model—has to be dynamically assessed, whereas the tutors' profiles may be predefined.
- *Teacher/tutor support for supervising individual exercises*. Essentially, this is a decision support function for the teacher. To achieve this, it is sufficient to aggregate the individual learner models in a form that allows filtering out of specific features, e.g. frequent problems. The support can also actively inform the teacher if desired.
- *Group formation with respect to given problems*. This is a generalisation of mediating peer help in that the number of group members is not restricted to two. Also, the problem requirements must be analytically specified.
- *Selection of adequate problems for a given group*. A problem is selected or generated in such a way that it could serve as a challenge to the group as a whole, but should still be feasible if the group were able to combine individual strengths.

In our description of a formal basis for these support functions, we follow the approach of "multiple student modelling" (Hoppe, 1995) and focus on the problem of intelligently mediated peer help. Based on a normalisation and standardisation of existing student modelling techniques, different types of student models can be analysed with respect to their formal characteristics for integration or "knowledge fusion". Overlay models are a simple case, since here we can ideally assume complete "additivity" (as a set union) of the individual portions of knowledge. If buggy versions are uniquely associated with correct rules, similar integration strategies can be defined. A general conceptual and formal framework for student model integration is introduced in Hoppe (1995) under the notion of *multiple student modelling*. The following two prototypical cases will be further discussed and formalised:

1. Given a number of students working on comparable problems in an open learning network, the system has to find pairs of students that could potentially benefit from each other in a joint session. The system supports the selection of a (human) co-learner. The selection criterion can be based on complementarity or on competitiveness, for instance.
2. Given a group of students, the system selects or generates a problem that forms an adequate challenge for the group as a whole. The problem should not be solvable on the basis of one student's knowledge alone, but it should be solvable on the basis of the union of the knowledge available in the group. In this case, the task for the group consists in knowledge integration.

These basic tasks can be formalised and standardised for overlay, buggy or perturbation models on the basis of general modelling primitives (cf. Figure 2) like *knows(Student, Topic)* or *has_difficulties_with(Student, Topic)*. Here, the predicate *knows(S, T)* denominates the general modelling hypothesis that student S knows a certain topic or dominates a certain skill T. It is also of interest to have explicit negative hypotheses: here labelled as *has_difficulties(S, T)*.

In a simple overlay type of model in which the complete set of knowledge elements K consists of elementary $k_i$ ($i = 1 \ldots n$), a student model is a sub-set of K. We may also relate certain sub-sets of K to "topics" (i.e. the knowledge elements required for this topic). For a student model SM in K and a (non-empty) topic T, we may say

*knows(S, T):< = > T is a sub-set of SM(S)*

and

*has_difficulties(S, T):< = > T without SM(S) is non-empty*

With buggy models, each student model is composed of correct (positive) and incorrect (negative) rules:

$$SM(S) = K^+ \cup K^-$$

Knowledge of a certain topic can now be expressed in the following way:

*knows(S, T):< = > T is a subset of the* $K^+$ *component of SM(S)*

and, accordingly, *has_difficulties(S, T)* holds if this is not the case.

Whereas the idealised fusion or integration of knowledge would just be calculated as the set union for two overlay models, additional assumptions are needed for the buggy case. Let us assume we have two students with buggy models:

$$SM(S_1 = K_1^+ \cup K_1^- \textit{ and } SM(S_2) = K_2^+ \cup K_2^-$$

Now, to predict if the two students should (ideally) be able to solve a given problem, we could postulate the following group student model:

$$SM(\{S_1, S_2\}) := (K_1^+ \cup K_2^+) \cup (K_1^- \cap K_2^-)$$

which is *not* just the set union, but assumes (1) that both student models cover the whole space of knowledge elements with either positive or negative rules and (2) that positive rules dominate or out-rule negative rules, so that only the intersection of the two "negative components" contributes to the negative component of the joint model. Of course, we do not know if this kind of knowledge complementation would really take place. It is the potential optimum based on the previously assessed knowledge states. Irrespective of the real performance of the pair of learners, this optimum can be used as a guideline for the selection of adequate problems or for defining scaffolding strategies that would favour the positive over the negative rules—for instance, in a dialogue game environment.

We will not further elaborate on the possibility of implementing modelling primitives for other standard modelling methods. However, given any implementation of these primitives, it is important that we can define cooperation criteria that can be used for the purpose of intelligently parameterising group learning situations, such as the criterion of *complementarity*:

*can_help(S1, S2, T) if knows(S1, T) & has_difficulties(S2, T)*

*Competitive* situations may be triggered by identical (or similar) profiles:

*competes(S1, S2) if for-all t in T: knows(S1, t) & knows(S2, t)*

It is evident that these definitions can also be formulated with fuzzy parameters or numerical estimates of the quality of the student's specific knowledge or skill. We may define the predicates *knows(S, T)* and *has_difficulties(S, T)* in terms of another specific predicate, *competence(S, T, Level)*, with *Level* being in a range of continuous or discrete values (e.g. "high" for *knows* and "low" for *has_difficulties*).

With these analytic means, we are also able to formulate a criterion for the selection of problems in such a way that they provide a challenge for a given group without being unrealistically difficult. Again, we assume the availability of individually assessed student models ($S_1 \ldots S_k$).We can now operationalise the selection criterion by saying an adequate problem should not be solvable by any one of the models alone, but it should be solvable by the union of all the models. This requires the predicate *solvable(P, S)* to be computable for any given student model, S and problem P. Computation is directly possible with executable student models, such as models, formulated as production rules or PROLOG clauses. For models that are not originally executable, solvability can still be operationalised by defining a precedence relation (*pre*) between the elements of a "knowledge space". For example, *pre(A, B)* would stand for *knowledge of A is a prerequisite for knowledge of B*. If we have a diagnostic procedure that gives us negative evidence (i.e. evidence that a concept is *not* known), we can now draw inferences of the following type: *given negative evidence for A we conclude that B is also not known*. Thus, we can extrapolate diagnostic evidence through a (directed acyclic) graph of the prerequisite relation. A more detailed model of knowledge spaces using AND/OR graphs has been introduced by Doignon and Falmagne (1985). It can serve as an elaborate basis to make competence predictions from student models that are not directly executable.

The integration or fusion of knowledge from several sources appears concretely in rules of the type *can_help(S1, S2, Topic)*, whose implementation makes use of more than one individual model. The management of several models at a time, including parallel dynamic update as a result of new diagnostic evidence as well as access to the models from a group monitor, requires a flexible modular architecture. For our diagnostic algorithms implemented in PROLOG, it is particularly desirable to be able to exploit backtracking over competing diagnostic hypotheses in parallel for all individual monitors. An architecture that complies with these requirements is described in Mühlenbrock, Tewissen and Hoppe (1997).

Hoppe (1995) exemplifies the parameterisation of group learning situations with a peer helper scenario around an interactive physics experiment. Here, the student model is a multifaceted competence model with mainly numerical facets. Modelling is cumulative for all learners in one central knowledge base and strictly sequential in one single thread, based on the processing of incoming notifications of user actions. Using the new multiprocess architecture mentioned above, we have also been able to implement multiple student modelling with a more sophisticated modelling method, namely *deductive diagnosis* (Hoppe, 1994). This technique is based on the meta-interpretation of PROLOG programs in very much the same way as described for the simulation model in section 3. The meta-interpreter is fed only with correct domain knowledge. Given an incorrect student answer, fail-safe interpretation yields error hypotheses that are disambiguated in a dialogue with the user or by adaptive testing. Disambiguation takes place in backtracking mode, i.e. we must be able to handle several diagnostic threads in parallel and to access the global group knowledge at the same time. This is achieved by a distributed PROLOG architecture with KQML-like basic communication operators.

Figure 3 shows a snapshot of a "teacher's assistant" in the mathematical domain of symbolic derivation. This interface allows for inspecting different features of the group modelling system based on inferences over individually assessed models (including the *can_help*, *knows*, and *has_difficulties* predicates).

The teacher's assistant is construed as a support tool in the hands of a teacher. It can be used for on-line supervision of exercises with networked learning groups, both in a "virtual" or in a presence setting. In this sense, the teacher's assistant is an add-on to an open interactive and collaborative learning environment. In our understanding,

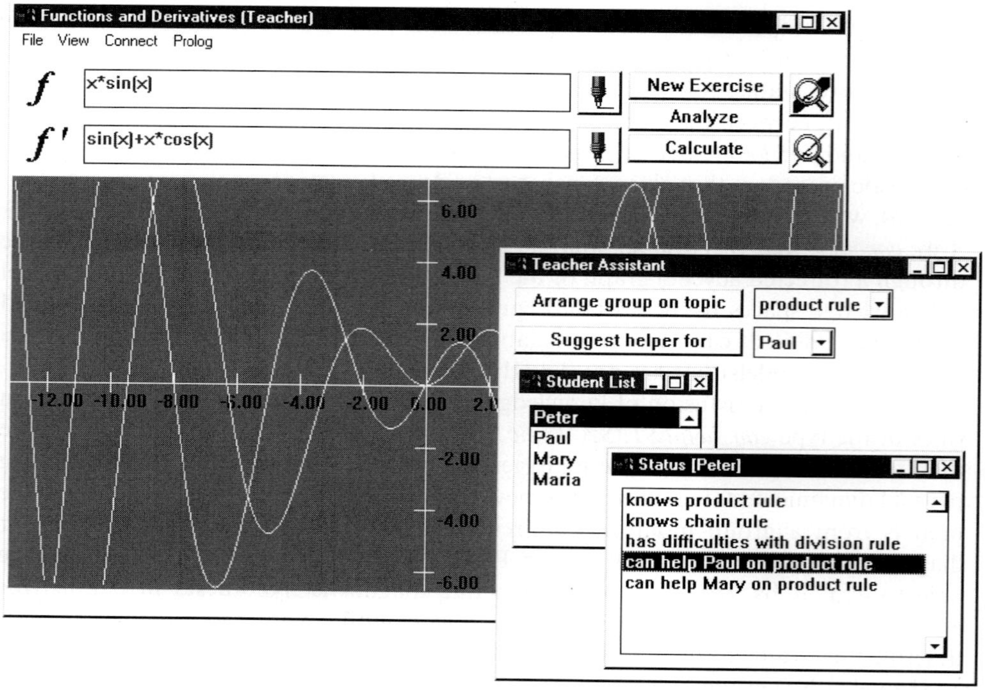


Figure 3: "Teacher's assistant" in the area of symbolic differentiation

intelligent support should not dominate or control the interaction structures in the environment. It is a background mechanism that should be of unobtrusive nature. Here, it can be consulted by the teacher, but there are also other similar mechanisms to support students directly: for instance, mechanisms to mediate and suggest peer help. The design of intelligent support functions must therefore be related to general considerations about designing collaborative learning environments.

## 5 Design Issues for Shared Workspace Environments

Intelligent support based on student modelling can be effective only if one considers what McCalla (1992) has named "implicit student modelling". An implicit student model, in this sense, is the set of assumptions and normative premises that guide the design of the learning system. Though implicit student models are an important element in the development of ILE, they are also relevant to ITS. With respect to group-oriented learning environments, we have to consider a number of design issues that are not relevant to single-user systems: in tele-presence systems, the types of communication supported and the channels through which communication is facilitated are basic design decisions. Technically enriched face-to-face communication, as it occurs in electronic or "computer-integrated classrooms" (cf. Hoppe, Baloian & Zhao, 1993), also is strongly influenced by design decisions concerning data sharing, turn-taking, control and initiative.

Our focus is on facilitating joint problem-centred activities in shared workspace environments with either natural face-to-face communication (as in a computer-integrated classroom) or technically mediated remote communication. Allowing only teletype dialogues may be useful for experimental studies, but usually one would prefer to provide richer forms of interaction. In open face-to-face situations, there is no need to restrict communication in parallel to shared workspace interaction. For remote communication, an audio channel is desirable. It is unclear, however, if full video-based tele-presence is a value-adding factor in problem-centred interaction, or if it may create too much distraction from the shared object space.

Concerning the type of collaborative or communicative activity, we focus on interaction or, more precisely, on the planning, coordination and execution of actions in the shared workspace environment. The general premise is that a shared manipulable external representation, dynamically co-constructed and modified, triggers coordinated cognitive processes. The effects of external representations on shared cognition can be manifold:

- The coordination of individual contributions is both constrained and mediated by the external environment (*coordination*).
- The individual contributions are given objective, material evidence in the external environment (*reification*).
- In addition to their content, the individual contributions may be illustrated by the external representation (*illustration*).
- The external environment may store individual contributions for later use, e.g. for the purpose of reflection (*storage*).
- The external environment relates different contributions to each other in an objective way ("*mise en relation*"; for a detailed elaboration on this Piagetian term, see Aebli, 1980).

From the point of view of analytic mechanisms, reification or "objectivisation" is of outstanding importance: observable actions correspond to changes of states in the environment. Essentially, the semantics of an action *is* this state change. In terms of chapter 3, we could say that expressions that take the form of an action in a shared external space are inherently *grounded*, so that, for analytic purposes, we avoid the difficult problem of disambiguating linguistic utterances. Roschelle and Teasley (1995) address this aspect in their qualitative analysis of collaborative learning episodes:

> "Since most of the utterances contain indexical, ambiguous references, the production of appropriate action both accepts and confirms a shared understanding of the task. Actions and gestures can likewise serve as presentations of new ideas." (Roschelle & Teasley, 1995, pp. 78–79)

They conclude that the external computerised representation can serve as a "*means for disambiguating language*" (p. 95).

We can distinguish two different modes of collaborative learning with respect to the role of language. In one mode, actions in the externally represented activity space primarily aim at supporting a linguistic discourse. In the second mode, which corresponds to our view, the external activity space is utilised as a kind of problem space, and external actions serve to solve a given problem. In both modes, physical actions and linguistic or speech acts are related to each other. However, in the first mode the overall goal is defined on the linguistic or discourse level and external actions are only means. In contrast, in the second mode the goal is defined in the external activity space, and speech acts play an instrumental role.

Of course, we do not want to suppress or restrict natural language communication in collaborative learning. Language-based negotiation is particularly important for shared planning and for evaluation. Shared workspaces "externalise" essential parts of the interaction in the environment, thus providing rich input material for automated analysis. The extent to, and the ease with which, communication and interaction can be externalised is a matter of design. In this sense, our design should support externalisation.

The implementation of shared workspace environments is still much more difficult than the implementation of interactive single-user systems. Therefore, it is particularly relevant to identify generic classes of manipulable external representations and to convert these into generic shared workspace systems. Currently, there is a special interest in *concept-mapping tools* and editors for *argumentation networks*. Concept maps form an external representation in which information is presented by structured graphs. The nodes in such a map represent concepts. The links between the nodes represent relations between the concepts. Once converted into synchronisable multi-user tools, they can directly support general collaborative activities, such as group discussions, or they can be specialised to provide more domain-specific visual representations.

Important applications of concept maps are the teaching of structural knowledge (e.g. Novak, 1990), the assessment of structural knowledge (e.g. Barenholz & Tamir, 1992) and the support of self-guided learning (e.g. Novak & Gowin, 1984). Concept maps support both the description of concepts relevant to the domain under scrutiny and the description of relationships between these concepts. If shared workspace environments are provided with protocolling functions at a certain level of detail, they are also an ideal tool for empirical studies of collaborative learning.

To support the collaborative construction of concept maps, we have developed a first prototype of a collaborative environment for the application domain of classical mechanics, as described in section 3 (Ploetzner et al., 1996b). The implementation makes use of a generic application for defining visual languages of different types in a distributed shared workspace framework. The design has been inspired by paper material that was used in a non-computerised investigation of collaborative problem solving and learning in the domain of classical mechanics (Ploetzner et al., 1996a).

The actions in the shared workspace are synchronised in tightly coupled mode (*WYSIWIS—What You See Is What I See*; for a detailed introduction to CSCW terminology, see Dewan & Choudhary, 1991; Greenberg et al., 1992). Collaborating partners can create and modify cards in the shared workspace in the same way as in their private workspace, without restrictions. Furthermore, it is possible to move cards from one workspace to another by a simple drag-and-drop procedure. A clipboard is not needed. Thus, parts of potentially different concept maps are interactively and collaboratively integrated during the cooperation phase. The need for integration is particularly motivated by the different concept sets available to the individual subjects. Since each subject has available a copy of the shared workspace, the final map or an intermediate map can always be stored and re-used collaboratively or individually. Figure 4 shows a drag-and-drop-action between a private and a shared workspace during a coupled session.

The environment was developed using the *MatchMaker library* for interactive multi-user interfaces. Synchronisation between shared visual workspaces is done object-wise through the broadcasting of user interface events from the direct receiver object to the set of coupled objects, thus triggering multiple execution in client applications. MatchMaker uses a client-server architecture with full replication of coupled objects. Initial consistency of coupled objects is by default, ensured through the over-

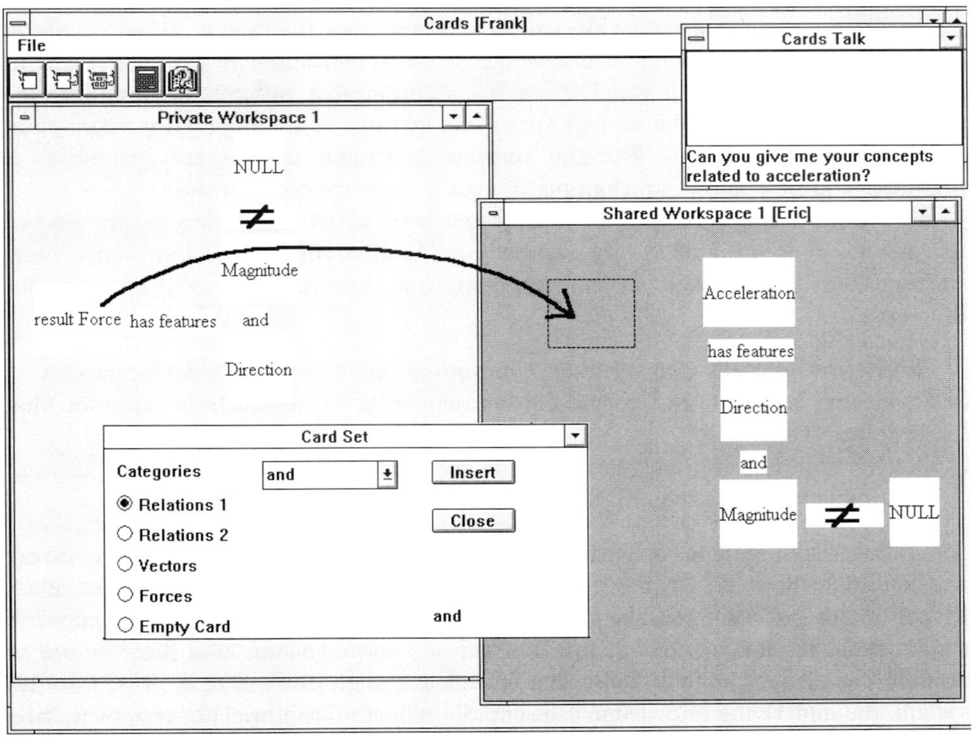


Figure 4: A drag-and-drop-action between a private and a shared workspace

writing of one object's content by that of another object (for more details, see Zhao & Hoppe, 1994).

In order further to support individual as well as collaborative problem solving, we also enriched the environment semantically with domain knowledge. The knowledge base is a static representation of domain concepts and their inter-relations. It does not represent user-specific or dynamic features. The knowledge representation features the following elements:

- terminological indexing of predefined concepts,
- set representation of algebraic equations in terms of the physics concepts that are algebraically related to each other, but without specifying the algebraic structure, and
- association of concepts with other concepts, including links between different types of concepts (e.g. the concept velocity may be associated with the concept acceleration, but there may also be links from the concept velocity to diagrammatic representations of linear motion).

The main emphasis of this approach is on an intermediate level of knowledge representation that is more than just lexical, but still below the level of a complete operational expert model. To some extent, it corresponds to knowledge about qualitative aspects of the application domain, only that there is a semi-analytic representation of algebraic equations in terms of the sets of concepts that they relate. What is specifically supported by this kind of knowledge is the construction of associative chains between related concepts in order to create a solution plan. Subjects working with the environment should be able to explore all the features of the underlying representation in a transparent way. For this purpose, a specific set of query primitives is provided. Figure 5 shows an example of such a supported exploration.

According to our view, support is different from active scaffolding on the part of the system. It is initiated by the subject's active querying of the knowledge base, starting from the concrete external representation. Typical types of queries are the following:

- Given a set of cards that represent concepts of any type, find related concepts.
- Given two concepts, find a relation between these or an algebraic equation that combines them.
- Given a solution, find inconsistencies or incomplete solution steps on the basis of semantic constraints (e.g. missing variable definitions).

Semantic enrichment aims at a crucial question of CSCL research: how can we respect and maintain the social or inter-individual dynamics and the principled open-endedness of group problem solving, but still provide substantial help in the domain? A similar point of view in which help is located on a more general level than the one of domain knowledge can be found in the Belvedere system (Suthers et al., 1995). In this system, the underlying knowledge is essentially related to rhetorical or argumentation structures in arbitrary domains of scientific discourse. Nevertheless, general constraints can be formulated: e.g. to detect circular argumentation or unsupported

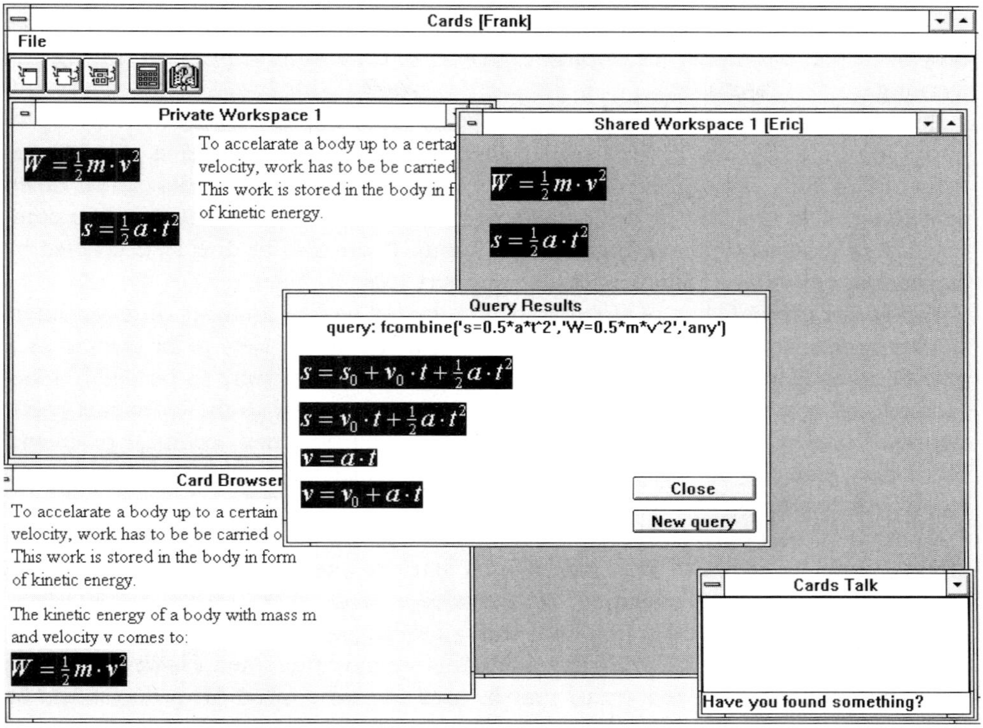


Figure 5: An example of supported exploration

arguments. In our scenario, since the activity is not discursive but constructive, we have to focus on domain aspects.

## 6 Conclusions and Perspectives

Designing collaborative learning systems puts analytic models in a rich context: models may be used implicitly, in the sense of McCalla (1992), or explicitly, in that they may be dynamically constructed and consulted at run-time. The role of analytic models is subordinate to overall design issues that range from epistemological principles over human factor considerations and aspects of communication design to technical aspects of system architecture.

We have exemplified the implicit use of a model in the design of our prototypical shared workspace environment for classical mechanics. In this environment, the design of complementary interactions, analogous to a jigsaw puzzles, reflects the basic mechanisms of knowledge complementation captured in the analytic model described in section 2. External representations derived from a cognitive model promise to be cognitively more appropriate for the posed problem solving and learning tasks. In addition, the use of shared external representations also provides an easy-to-inspect data space for empirical analyses of collaborative problem solving and learning.

Multiple student modelling is a first step towards explicit student modelling in group learning. In our current implementations, the group model is constructed from individually assessed models of single learners. The group model can be taken advantage of at run-time to make "informed" decisions with respect to group composition or group-adaptive selection of problems. It can also be directly consulted by teachers or tutors as a didactic decision support tool.

Given our current state of technology, automated on-line analyses of group interactions appear to be practically feasible either if the group were to be treated as a whole—as suggested by Paiva (1997)—or if the interactions were to be highly schematised and constrained to be able to assign credits correctly to the individual participants. Treating the group as a whole does not even pose new technical problems, but it may give rise to interesting new questions, particularly if the uniform group model was combined with previously assessed individual models.

Unrealistic assumptions about the predictive power of modelling learners in a group should be avoided. It is indeed very hard to predict whether collaboration and the mutual complementing of knowledge and skills really takes place. Practically relevant success conditions, such as affective, social and ecological conditions, are not covered in the analytic models that we have presented. However, models of learners collaborating in a group may be used to predict what the group *should* be able to achieve, just based on the given cognitive competencies. That is, the use of such models would be *normative*! Normative predictions can be used to provide adaptive support for making the desired and theoretically possible group interactions occur or, at least, to estimate the "under-achievement" of the group.

Concerning the semantic enrichment of shared workspace environments, it is still an open issue how exactly intelligent support will coexist or interfere with human–human collaboration. In our mechanics example, semantic support functions are passive, i.e. there is no intervention on the part of the system. It is a subject for further research to investigate whether and how often such intelligent resources are actually used. It will be particularly interesting to see to what extent system support may be replaced by group interaction, and vice versa.

## Acknowledgements

This research has been supported, in part, by the German National Research Foundation (DFG) under contract "Pl 224/2–2" and, in part, by the European Science Foundation (ESF) within the programme "Learning in Humans and Machines".

# 9

# Using Telematics for Collaborative Knowledge Construction

Tia Hansen, Lone Dirckinck-Holmfeld, Robert Lewis and Jože Rugelj

When engaging in collaborative knowledge construction, geographically distributed groups are found to have additional problems in comparison with groups working face-to-face. Telematic services have been created to address these problems, but so far they have not resolved all the difficulties. One of several reasons may be that too little consideration is given to the match between the tool and the process it is intended to support. Thus, this chapter first describes some functionalities of telematic tools, then offers two typical present-day examples of collaborative knowledge construction by groups that use such tools—one is a case of project pedagogy distance learning, the other is a case of international research collaboration. A promising framework for understanding collaborative and tool-mediated human effort is the dialectically systemic approach of "*activity theory*", which originated with Vygotsky's cultural–historical school of psychology in the 1920s, and is presently being further developed by Yrjö Engeström, amongst others. Aspects of this theory are outlined and used as a framework to further the understanding of some processes and problems that were reported on the basis of the case studies.

## 1 Introduction

Collaborative knowledge construction (CKC) has provoked extensive study in recent years. For a host of real-world problems to be solved, several different areas of expertise are required, and this, in turn, requires the participation of a group of people.

Coordinating different people's concerns and expertise has never been a straightforward task, and, in reality, the added burdens of communication and coordination sometimes outweigh the productivity gains that, in principle, should be possible when combining different people's knowledge (e.g. Galegher & Kraut, 1990).

Additionally, the people involved in CKC may not be able to meet face-to-face in order to coordinate their ideas and expertise. For this additional problem, the research and development domain of computer-supported cooperative work (CSCW) offers solutions in the form of telematic tools. These tools enable communication between people who are dispersed in space or who participate at different times. As such, the tools are potentially useful for a wide range of people. However, in this chapter, the focus is on groups for whom knowledge construction is the primary aim of their collaboration.

There are two main categories of groups for whom this is the case. The first category is workers whose job is to produce knowledge not hitherto available in society. Researchers are a prototypical example. The other category is that of learners, i.e. people who (re)construct knowledge for themselves that others may already hold.

It can be argued that there are principled differences between workers and learners. What will be emphasised here, however, are the similarities of the processes in which both are engaged. Though the knowledge that learners construct is not new from the perspective of a wider community, for the participants it *is* viewed as new.[1] Thus, it can be claimed that when workers and learners try to construct knowledge collaboratively, they are engaged in somewhat similar processes and, when the group is dispersed, they will both need the same kind of tools to support their communication.

During the collaborative construction of knowledge, geographically dispersed groups are known to have more problems than groups who are able to work face-to-face. Telematic tools have been created and used to overcome these additional problems and some of them provide clear benefits. However, some problems persist; dispersed groups using telematic tools for collaborative knowledge construction still report or display problems that exceed those found when knowledge construction can be managed face-to-face (Dirckinck-Holmfeld, 1990).

At least four factors contribute to this relative lack of success:

1. hardware limitations,
2. software limitations,
3. user resistance and
4. lack of ability to choose and apply the right tool for a given task.

The hardware limitation factor refers to constraints in speed, bandwidth, etc., of the technology available at present and should diminish with the technological progress over time. Software development is also progressing from primitive to more complex applications. Of course, software can only be as advanced as the underlying hardware permits, but software limitations are not fully reducible to hardware limitations. Another problem for software designers is that in order to develop an effective tool, a deep insight of what it should support is necessary. Software designers sometimes complain that they do not have a sufficiently good understanding of why something does not work. The associated methodologies of investigation, and the underlying conceptions of human cognition and collaboration, need to be sophisticated so that designers can be guided more effectively (Nardi, 1996; Rugelj, 1996, 1997). User resistance is another issue to which psychodynamic, organisational, political, and a host of other factors can contribute. However, resistance from users can also be related to technological and design factors. Primitive interfaces, complicated procedures, and lack of some of the features that are crucial for carrying out the tasks cannot be expected to convince people that investing effort in learning to use a tool is worthwhile. A fourth kind of problem may be traced to a lack of ability to choose and

1. The resemblance between learners' and researchers' processes may be slight in very formal, teacher- and fact-focused learning, but becomes pervasive in pedagogy that emphasises students' active efforts at finding possible answers to open, real-life problems.

apply the right tool for a given task. In other words, we believe that there is a lack of knowledge—as well as of awareness—concerning which types of communication and collaboration tasks may best be supported by different telematic tools. It is this latter factor that will be discussed in this chapter.

With this purpose, we first provide a general perspective on telematic tools, in order to give a context for their potential use. Some roles and functions of such tools are described and mapped onto needs and tool-related problems that will be analysed in more depth later in the chapter.

The subsequent section summarises some findings from two naturalistic case studies: one concerns project pedagogy as a methodology for computer-supported collaborative learning (CSCL) at Aalborg University (Dirckinck-Holmfeld, 1990), and the other concerns computer-supported cooperative work (CSCW) within an international, cross-disciplinary group of researchers (VMDL, 1996). These studies identified partial tasks and processes that differed in the extent to which the support from the available telematic tools seemed adequate.

The chapter continues by suggesting an approach to conceptualising of collaborative knowledge construction, inspired by the cultural–historical school of psychology and its offspring "activity theory". This paradigm sees motivated activeness, tool usage, and collaboration as fundamental characteristics of human effort. It therefore provides a framework within which one can reach more theoretically grounded conceptions of human endeavour and allow integration of fragmented, partial understandings within certain fields (e.g. the field of user-interface design; Nardi, 1996).

The final section applies the activity theory framework in revisiting some problems reported by the case studies, now embedding them in a more principled discussion of CKC processes and their requirements with respect to the participants and the tools that they use.

## 2 Tools to Support Collaboration

The term "telematics service" has a very broad meaning, since it can be used to designate any integration of computer-supported information processing and communication. Various sub-sets of telematics services can be found in the literature under any of the following names: computer-mediated communication (CMC) tools, groupware, teamware, tele-conferencing-systems, electronic meeting systems, group decision support systems or computer supported cooperative work tools.

Shared knowledge is fundamentally constructed through collaboration between persons, or within groups, involving the creation and interpretation of communication. Different forms of interaction—which can be extended in time and space by various media and telematics tools—are the means by which people both collaboratively construct beliefs and meanings, and state differences between them. Interactions require and create a *common ground* or shared knowledge space of beliefs and assumptions during conversation (Clark & Brennan, 1991; see also chapter 3, this volume). Therefore, one of the major issues facing designers of systems for the support of collaborative knowledge construction is how to help members of a group to understand each other and to create and maintain their common ground.

Systems for the support of collaborative knowledge construction relate functional features with social aspects of team work. Each functionality has an impact on work behaviour and the efficiency of the whole group using the system. Furthermore, the functionality of such a system also influences the behaviour of individual group members. However, psychological, social and cultural processes within groups of collaborating people are the key factors in determining whether systems for CKC support will be accepted and successful (Reinhard et al., 1994). In section 4, we introduce a framework for considering such factors.

### *2.1 Telematics Support for Communication*

Collaboration that is mediated by telematics may proceed *synchronously* or *asynchronously*. In synchronous interaction, participants can be geographically dispersed but are temporally copresent, whereas in asynchronous interaction, collaborators' use of the tool may be dispersed in time as well as in space. These two forms of interaction each have their merits, e.g. synchronous interaction allows participants continuously to align their understandings, whereas asynchronous interaction allows subordination of the collaboration to diverse working patterns (availability) of participants.

Another useful distinction concerns *implicit* versus *explicit* communication. From a designer's perspective, the former refers to collaboration through use of shared information resources, such as documents, images and spreadsheets, whilst the latter is explicit communication among collaborators using audio and/or video channels, or just simple text messages.

Coordination of interactions deals with implicit as well as with explicit interaction within the group of users. Coordination of implicit interactions includes the provision of consistent external representations of shared information. Coordination of explicit interactions depends on the group size and on the group's task. The need for coordination grows with the size of a group, the diversity of members and the complexity of task. A survey of the mechanisms for coordinating interaction can be found in Watabe et al. (1990).

Users of systems for the support of collaborative knowledge construction must select one or more types of communication media in order to interact with other members of the group. Cockburn and Greenberg (1993) have identified several factors that affect the selection of an appropriate medium:

- characteristics of a task,
- availability of telematics tools to all members of the group,
- intended period of interaction,
- inertia involved in switching to more appropriate media,
- the ability to extend interaction in time and space.

An important dimension of a communication channel is what has been termed its *richness* (Daft & Lengel, 1984). This can be thought of as the potential information-carrying capacity of data. Factors that contribute to richness are suggested to be:

- interactivity (the speed of reaction),
- multiple cues (verbal, intonation, proxemic, and kinetic),
- language variety (numbers, natural language, symbols, images),
- socio-emotional cues (social presence, feelings).

A recent research project (VMDL, 1996) and the interpretation of richness discussed by Heeren and Lewis (1997) suggest that these factors are made explicit in the way that people who are working collaboratively express their perceptions of the various channels at their disposal. Tables 1 and 2 reproduce the VMDL project's overview of advantages (and disadvantages) offered by different media and their combinations. Some implications of member diversity and task complexity are discussed more deeply in relation to the case studies (sections 3 and 5).

### 2.2 *Transparency and Roles of Tools*

The question of the tool's role can be thought of from two analytically distinguishable points of view: one can think in terms of the tool as compensating for difficulties, or in terms of the tool as facilitating new processes.

Tools can *facilitate* human thought through appropriate processing and presentation of data in such a form that is easier to understand. The idea of facilitation is very close to Hutchins'' (1990) concept of tools as "re-representations", i.e. alternative ways to show the same information. With increasing processing power and multimedia input/output devices, facilitation tools are becoming more and more efficient. In other words, they enable users to see what could hardly be imagined before the invention of the tool. Thus, they allow users to do more than before, with the same data.

From the *compensation* perspective, the tools alleviate difficulties that users have in communication and collaboration due to distribution of collaborating partners in time and space. According to some findings (Lubich, 1995), tools can best serve a compensatory purpose when they are *transparent* to the user, in the sense that the communication supported by these tools is, to a great extent, similar to face-to-face communication. This attribute (proposed by Ihde, 1975) can be used for classification of tools according to their degree of transparency.

The degree of transparency of tools (or of the user–tool relationship) refers to whether (and to what extent) the tools become an extension of the human users or remain separate from them in their perception.

Ihde (1975) has discussed the type of human–tool interaction and the degree of transparency of tools. He argued that there are three types of interaction according to different levels of transparency. The first level in his analysis occurs when the human is using an implement in which the tool becomes transparent. At this level, the tool operates as an extension of the user's body, and the user can feel the environment

Table 1: Advantages and disadvantages of media

| Medium | Advantages | Disadvantages |
|---|---|---|
| Open computer conferencing | • The communication is stored<br>• Communication in your own time, when it is convenient to you<br>• Time for reflection<br>• Flexible structuring<br>• Communication less influenced by formal roles | • Difficult for decision making, especially a result of unclear meaning of non-responses<br>• Large volume of communication, difficult to intervene for those not fully involved<br>• Daily expectation of focus not appropriate for those with many obligations |
| Closed computer conferencing | • Limited audience, increased trust, and hence more openness<br>• Useful for emotional support<br>• Less volume of communication for those outside the small group | • May make other members feel excluded, thus distorting the social relationships within the community as a whole<br>• May lead to more attention than necessary to social issues, rather than to the work to be carried out |
| Formal face-to-face meetings | • Useful for focusing on the team task<br>• Useful for making decisions<br>• Useful for getting to know one another and building a sense of community | • Involves high cost and use of time<br>• Onerous and/or imperfect record of debate |
| *Ad hoc* face-to-face meetings | • Useful for developing shared understanding<br>• Useful for generating new ideas together | • Only those that are co-located and have time can participate<br>• Increased sense of distance for those who cannot participate<br>• Danger of not sharing outcomes with the rest of the team |
| Electronic mail | • Useful for tackling urgent issues<br>• Useful for one-to-one conflict solving | • E-mail messages may get "lost" in the dozens of messages people receive daily |
| Fax | • Useful for easy distribution of graphics and tables<br>• Useful for drawing attention to urgent issues | • Cannot be edited or stored electronically |
| Telephone | • Useful for spontaneous exchange<br>• Useful for resolving conflict | • Not usually recorded or storable for later elicitation |

Table 2: Relationships between usages of different media

| Related media | Character of relationship |
|---|---|
| Open computer conferencing and formal face-to-face meetings | Communication in the computer conference forms an important basis for decision making face-to-face |
| | Face-to-face meetings can be used to plan the further work carried out through the computer conference, and decide on the new item structure for the computer conference |
| | Face-to-face meetings may be used to establish criteria on which to make decisions in the computer conference |
| Formal and *ad hoc* face-to-face meetings and computer conferencing | Face-to-face meetings can be useful for efficiently developing shared understandings that could be built upon in the computer conferencing |
| Fax and computer conferencing or e-mail | Fax may be used to share graphics and tables that were part of texts, such as report drafts, but that were more complex to distribute through computer conferencing or e-mail |

directly through the tool. With the second type, direct tactile experience is lost; a level of translucency is reached here, as the tool still extends the individual's perceptions, but in a reduced form. In both of these types of interaction, the tool can disappear into the background, not requiring attention in itself. In the third type of relationship, this extension of bodily experience is lost, the tool becomes opaque and separate from the user and acquires its own identity in the user's perception.

We suggest that "compensating" and "facilitating" roles have different requirements in this classification. When compensating, the tools should preferably be transparent so as not to "steal" the user's attention for themselves. This is not so necessary for facilitation tools.

Interestingly, there is a new and rapidly growing group of tools that seem to switch their role. These tools are designed with facilitation in mind, but integrate, at the same time, facilities for communication and collaboration support. Thus, they also compensate well for distance between collaborating users. Typical examples are collaborative drawing tools and editors, whiteboards (Aytes, 1995), and multi-user virtual environments.

In the past, conferencing tools were asynchronous and text based (as were all other computer–user interfaces). Nowadays, the term "conferencing" is broader and comprises also synchronous audio- and video-conferencing tools, which can be implemented on standardised workstations (Bennett & Kirstein, 1997) and even personal computers (Dorcey, 1995). Such desktop conferencing tools may replace face-to-face meetings fairly well, at least when the interaction concerns more formal activities.

However, in everyday situations of collaboration over distance, such multimedia conferencing is still rarely an option. At present, the main problem for application of these tools in "real-life" situations of computer-supported working and learning is that they require a greater capacity than most communication networks can handle at present (predominantly a hardware limitation factor). Thus, whilst experimental and quasi-experimental settings have provided a great deal of knowledge about the usefulness of such tools (e.g. Rugelj, 1992, 1996), it has not yet been possible to carry out large naturalistic studies of cases in which these tools play a major role. In the cases we describe next, the conferencing facilities were of the "classical" type.

## 3 Two Case Studies in Collaborative Knowledge Construction

Following the current perspective that knowledge needs context, this section provides a context for a more theoretical analysis of the complexity of collaborative knowledge construction. We offer two examples: case 1 involves CSCL, and case 2 involves CSCW. In both cases, the objective of the collaboration is knowledge construction, and both represent typical present day activities for this purpose. The descriptions in this section are based on findings and interpretations from two naturalistic research projects, carried out by Dirckinck-Holmfeld (1990), and by Lewis and colleagues (VMDL, 1996), respectively. In section 4, we suggest a theoretical framework for further analysis of problems such as those described in the cases.

### *3.1 Case 1: Problem-Oriented Project Pedagogy as CSCL*

Aalborg University offers several distance education courses. Besides two on-campus weekends per semester, a "classical" computer conference system is provided for communication between students, between students and teachers, and between teachers (several part-time teachers from the professional sector). Teaching is organised around modules, each of which has its own conference on the system, and in the form of projects performed by groups of students.

All courses at Aalborg University have been designed around a methodology called "problem-oriented project pedagogy" (PPP). This form of pedagogy employs case-based learning and problem-solving learning, but must be distinguished from traditional versions of both of these paradigms, because it holds different assumptions about what constitutes active acquisition of knowledge. In problem-oriented project pedagogy, it is not the problem solving *as such* that is important, but the general problem orientation, and the fact that the students themselves must perform a problem setting with respect to an unresolved problem or a phenomenon that is not fully understood. Critical reflection on a (scientific) problem or a phenomenon in society is the basic didactic principle. The emphasis on critical reflection, and the development of the ability to *formulate* (set) problems, both contribute to the fact that problem-oriented project pedagogy is a more demanding form of learning than approaches that focus on solving a *given* problem (Fjuk & Dirckinck-Holmfeld, 1996).

Furthermore, according to Illeris (1981), problem orientation does not, in itself, constitute the foundation of an active process of acquiring knowledge through critical reflection. An additional condition must be met: "A problem is not a problem in a psychological sense if the person who has to work with it does not experience it as a problem" (Illeris, 1981, p. 83, our translation). Creativity, commitment and motivation are crucial aspects in relation to critical reflection. When the students themselves define and formulate the problem, they have a conscious relation of "ownership" with respect to it and are inherently involved and motivated. Illeris refers to this as "participant control".

Thus, in this pedagogical paradigm, participant control and problem orientation are interdependent and constitute the framework of the acquisition of knowledge. These didactic principles are implemented using collaboration organised in groups. A typical project is of several months' duration and constitutes around half of the students' study time during this period. This organisation builds on the principle that understanding is developed in situations where cognition, experience and attitudes are confronted and weighed against each other. Apart from the individual acquisition of knowledge, the concrete result of the collaboration is a collaboratively produced project report.

Project pedagogy has been the pedagogical model for Aalborg University and Roskilde University in Denmark since they both started in the 1970s. Since 1987, the model has also been the conceptual foundation for computer-supported distance learning at Aalborg University. It began with naturalistic experiments (Dirckinck-Holmfeld, 1990) and from 1991 ran full-scale with a constantly increasing number of courses: humanistic informatics, human-centred technology, health informatics, geo-informatics, management of technology, computer science, and German culture and language. Since 1987, research and developmental work has gone on more and less continuously and is documented in several reports (Dirckinck-Holmfeld, 1990; Georgsen, 1995; Fjuk & Dirckinck-Holmfeld, 1997).

In what follows, we will describe some of the tool-related problems with CKC within computer-mediated problem-oriented project pedagogy in geographically dispersed groups.

### 3.1.1 Communication Patterns during Phases of CKC

A detailed analysis of two project groups' use of the conferencing system was carried out by Dirckinck-Holmfeld (1990).[2] One of the groups carried out a genuinely collaborative project, with a shared focus, specified problem-setting, integrated research

2. To evaluate these findings against newer research, it should be noted that the data is from Denmark's first large-scale naturalistic experiment with computer-supported collaborative distance learning, performed in 1987–1989, and that two factors related to this condition significantly influenced the communication pattern: (1) the students were provided with computers by a company at no cost to the students and their telephone expenses (transmission costs) were paid for by the project, and (2) despite good technical support for the conference system (PortaCOM) there was a high number of technical problems.

process, etc. The other group was a semi-collaborative group that wrote a project together; their problem formulation being sort of an "umbrella" within which they wrote a project with a strong division of labour, with participants each writing their own sections. Dirckinck-Holmfeld (1990) compares the two groups' communication patterns, but here only findings from the genuinely collaborative group will be discussed.

The students displayed different communication patterns during different phases of the project, as shown in the following sub-sections. (A more principled account of phases in CKC can be found in section 5.3).

### 3.1.2 Group Formation and Problem-Setting Phase

During the initial phase, many ideas—concerning who was to be in the group, how they were going to work, and what to write about—were sent to the conference. However, analysis of the content of the communication and the complexity of the interaction revealed that truly interactive sessions were infrequent. Most of the communicative acts were proposals and questions sent to the conference, but they elicited few comments and no real discussion amongst participants. The problem-setting phase proceeded for more than 2 months and the students expressed much frustration because they did not feel that any progress was being made. The problem-setting phase was not completed until eventually the students arranged a private weekend meeting face-to-face.

### 3.1.3 Planning and Research Phase

During this phase, a somewhat different pattern was visible. The content of the communication was social issues, technical issues, planning issues, and discussions concerning the actual subject matter of the project. The technical issues took up much time and space in the communication, sometimes up to 20%. This is a large proportion for something that is supposed to help, rather than being an object of attention in itself (cf. the discussion of transparency needs in section 2.2). Another interesting finding from the communication on technical issues was that—in comparison with the communication for problem setting—the interactivity was a little higher. For planning issues, the same pattern was found; there was considerable planning communication throughout the whole period (25%), and communicative acts concerning this topic were rather interactive. The rest of the communication dealt with social issues or the subject matter of the project, with most communicative acts being on the latter. As opposed to the problem-setting phase, in the research phase the communication became increasingly interactive throughout the period, with students commenting on each other's drafts.

### 3.1.4 Conclusion Phase and Evaluation

The last phase of the project was the most interactive period, and the communication showed lots of energy. The students sometimes "met" on the conference system during a specific period in the evening to discuss certain problems in the contributions. The group ended their work by giving the project report a final revision during a private face-to-face weekend.

This group wanted to write a fully integrated project in which all participants worked on the same problem, and contributed equally to it. Their project work, as such, was relatively successful. In comparison with on-campus groups, on the one hand, and distance groups without support from telematic tools, on the other, the result produced by this group seemed to fall in between the two. When compared with experiences from ordinary face-to-face project learning on campus at Aalborg University, neither the communication nor the resulting project report were nearly as interactive, critical and focused as is usually the case. On the other hand, when compared with project-based distance learning that is not supported by telematic services, this was a much more integrated project (see also Dirckinck-Holmfeld, 1990).

Thus, in order to help the development of a critical and truly collaborative process and product of problem-oriented project learning as distance learning, a deeper understanding of the collaborative learning and communication process is needed. This also requires matching the highlighted needs with new tools and practices (see also Fjuk & Dirckinck-Holmfeld, 1997).

#### 3.1.5 Summary

For some functions (technical issues, planning in intermediary phases), conferencing seemed adequate in enabling interactivity. However, for other functions (in particular, problem setting, but perhaps also synthesis/conclusion), it clearly was not. Problem setting is a notoriously difficult phase, even for face-to-face groups; but for distributed groups, it seems to be nearly impossible, despite support from telematic tools that work well in other phases. Referring to section 2.1, task complexity seems to be the main issue here. However, diversity of members in the group may also play a role; for instance, groups that have previously worked together, or for other reasons have a large "common ground" at the outset, may find it easier to overcome the problems caused by task complexity than groups that have not. These factors are discussed further in sections 4 and 5.

### *3.2 Case 2: Project Research as CSCW*

For research in computer-supported cooperative work, concern with real-world "open" problems as the object of collaboration is quite natural. In addition, this field has traditionally intersected with human–computer interaction research (i.e. user-interface design and issues related to this), and has discussed the various roles, functions and phases in a collaborative process, together with the interdependencies between these and requirements of the telematic tools. Thus, the case of collaboration in an international research group, presented in the following sub-sections, elaborates on some of the questions raised in the previous case study.

### 3.2.1 Task Outline

The European project group described here was one of a number of distributed research communities studied in the VMDL project (VMDL, 1996). It was a relatively small (30 members), international, tightly planned, short-lifespan (17 months) research community using several different communication channels, including face-to-face meetings and electronic media.

The majority of the researchers involved were academics who participated in the project on a part-time basis, besides fulfilling their usual administrative, teaching, and/or other research tasks. They were also members of a much larger EU community concerned with students' mobility and credit transfer. The larger community had been engaged, for a number of years, in finding ways of ensuring that students were able to make satisfactory choices in following course modules in other European universities. The way to achieve this was by the careful preparation of course descriptions, but such a mechanism was far from satisfactory.

The project set out to experiment with a technology-supported solution to this problem, which required the establishment of a common, shared language for describing the competencies that were learned as students followed particular course modules. The "language" had to enable effective pan-European communication between educational institutions about a particular subject domain.

The community's work involved the following global activities:

- the development of a specific "language" for describing an educational subject domain,
- the creation of specific, locally valid descriptions of parts of the subject domain using this "language",
- the pooling of the descriptions in a common database,
- the negotiation of these descriptions among the sites, aimed at minimising the number of different descriptions,
- the pilot implementation of these descriptions in their target sites.

### 3.2.2 Motives for Participation

Members' individual motivations were partly content-related. They often expressed a personal concern with the idea of common "language" development as they were committed to improving opportunities for student mobility. Many members had also career-related motivations, especially those who had short working experience. Others considered the project and their community membership as just part of their working life, or were asked to join and contribute by their colleagues.

Members carried with them the motivations of their parent organisations, which were related to the content of the community's work, but also to prestige, money, and extension of the network of relationships, all of which are beneficial for continuation and growth of the organisation in the medium and long term.

### 3.2.3 Communication and Collaboration Patterns

Computer conferencing was the main communication channel used, and there were occasional face-to-face workshops. A common database of course module representations was created during the project. Members wrote their course module descriptions in local databases, sent their data to the central site, and received updated copies of the common database.

Besides this whole-group communication, local communication and collaboration took place among researchers at each partner site. Members from one site prepared their contributions to the computer conference collaboratively, and then submitted them to the whole group as one "person". In addition, electronic mail, fax, and telephone were occasionally used.

After each face-to-face workshop, the computer conference was given a new structure to account for the new set of tasks that were decided on for the following period. This restructuring also gave impetus and new energy to the conference. The old conference items were archived but remained accessible for reading throughout the whole project. Referring to the discussion of coordinating "implicit communication" (section 2.1), updating conference structure to reflect current task structure can be seen as an instance of ensuring shared external representation.

### 3.2.4 Division of Roles and Responsibilities

The community's work was structured around nine work-packages, each with a work-package leader. There was one general project coordinator. The work-packages were planned in advance, and delineated by deadlines and dates of deliverables.

Within each work-package, the work-package leader took formal responsibility. This included planning, meeting deadlines, and initiating and coordinating the writing of deliverables. In the beginning of each work-package, the members developed and decided collaboratively how to approach the work-package task.

In this project group, leadership was a complicated issue, since giving someone authority is, in itself, a decision to be made. Furthermore, some participants may consider authority over peers incompatible with democratic ethics, some may prefer decisions to be made by implicit consensus, etc. A solution appeared by taking turns in moderating the discussion and decision making.

Conference item moderation was central to the community's success. In the first stage of the project, the leader of the current work-package occasionally put summaries and opinions in the conference, but did not draw conclusions. Members from other sites did not do so either, because they felt they did not have the "authority" to do so. As a result, the discussion went in circles. During one face-to-face meeting, it was decided to remedy the problem by using a moderator, and the work-package leader was urged to take this role. This worked much better, since she subsequently closed discussions by making decisions based on reviews of the arguments put forward during the discussion.

During the later stage of negotiation of course module descriptions, a new decision-making procedure was implemented: each site took turns in taking the role of moderator. This also ensured more "ownership" of the material in all the partner institutes. Each site made an agenda, led the discussions for 1 week, reviewed what was said, and made a decision. Then, another site took over the moderator role.

### 3.2.5 Decision Making

Many decisions concerning the overall structure, management, and tasks of the project and the community were embedded in the project proposal. During the project period, fine-tuning of activities frequently required additional decision making. Such decisions were mostly made collaboratively during face-to-face meetings, with the work-package leader taking a leading role, but within the framework of a "democratic" process.

A problem did occur when some members thought agreement was reached, whilst others thought this was not the case. To resolve this, a great deal of debate and discussion followed in the next face-to-face meeting. The decision to have moderation was made in one of the later face-to-face meetings after considerable confusion via e-mail and computer conferencing.

Decision making whilst communicating through a computer conference appeared to be difficult, but possible. The most important requirements were that a procedure ("rules") for making the specific type of decision had been established, and that someone (such as the moderator) had taken the leading role.

A main benefit of meeting face-to-face was that the "rich" interaction helped the team to adhere socially, as well as to develop a common understanding of the tasks and of the ways to make progress.

The combination of face-to-face meetings and computer conferencing was thought to work perfectly well—if key structural elements were set: namely, a good moderator or chair, and explicitly stated rules and roles. One participant summarised the relationship between the two media as follows: "Face-to-face meetings were used to decide on 'the rules of the game', the computer conference for 'playing the game according to the rules'."

### 3.2.6 Summary

This case is complementary to the former one in a number of respects. It was a CSCW as opposed to a CSCL group. The goals of the collaboration were defined in the original project proposal rather than being negotiated within the group. The period of collaboration was longer, and it was a much larger group than the one described in the CSCL example. With a group of this size, there were variations in members' motivations, issues of leadership were important and it was necessary to define roles within the group.

However, there are also similarities between the two cases. In both, complex tasks such as decision making seemed to be hard to accomplish by computer conferencing, and ambiguity reigned until a face-to-face meeting had been held. On the other hand, once key decisions had been made, conferencing allowed participants to proceed.

There will be further examination of these case studies in later sections of this chapter. First, it is necessary to present a theoretical framework that can assist this analysis.

## 4 Understanding Collaborative Knowledge Construction

One of the most promising frameworks for understanding collaborative knowledge construction within a systemic approach is that of cultural–historical psychology/activity theory (CHAT). CHAT sees the dynamics of human activities as the integration of various interdependent relations in a dialectical process, where mental and external processes continuously force each other forwards.

The basic conceptions of human activity as systemic and dialectical originated from the cultural–historical school of psychology that was founded by the Soviet psychologist Vygotsky in the 1920s, and was further developed in the 1930s in close collaboration with his students Luria and Leontiev, whose version is known as "activity theory". The basics of Vygotsky's conceptions have spread into several lines of theory and research, of which one of the latest applications is to human–computer interaction research. In this case, CHAT promises to provide a framework for understanding the systems that computer-use supports (Nardi, 1996). The inspiration for the analysis in this chapter primarily follows interpretations presented by Engeström (1987), Cole and Engeström (1993), and Nardi (1996).

One fundamental assertion of CHAT is that humans are distinguished from other animals by the way in which tools are used. Human activity is seen as inherently mediated by tools, which change not only the physical conditions of life but also mental processes. The way in which tools are produced, kept, handed over and used results in a development of the structures and processes of human cognition that is not purely determined by current environmental and biological conditions, but also by the interaction with other humans and the cultural environment inherited from the past.

A person's interaction with the environment is partly structured by the set of tools that has accumulated during history and that, in a certain sense, can be said to *be* the culture into which human individuals are born and within which they develop. Thus, human thought and behaviour is said to be culturally mediated via the tools that individuals appropriate for supporting and structuring their own thinking and behaviour (Leontiev, 1978; Vygotsky, 1978; see also Säljö, 1996, for a general introduction to this view of tools as cultural artefacts to be appropriated by the individual).

Here, the term "tools" is used generically for objects and other phenomena that can externalise accumulated human experiences. These range from specific material instruments (such as hammers) to general semiotic instruments (such as language). Tools must be understood as providing, at the same time, possibilities and constraints on the actions and perceptions of the persons using them. They provide possibilities because they provide access to historically accumulated experience in condensed form, and constraints because by setting the one perspective that is condensed in the tool, other perspectives are rendered less accessible (Bygholm & Dirckinck-Holmfeld, 1997).

Since human activity is always mediated by tools, and tools provide access to knowledge gained by predecessors, other people are always indirectly present in a person's activity towards some object. Mostly, other people are also directly present in the system as collaborators or as part of the object. Thus, the minimal structure of human activity is a system comprising at least the following: subject, tool, object, one or more "other subjects"—and relations between all of these (Mammen, 1989; Cole & Engeström, 1993).

From the perspective of CHAT, complex human activities cannot be fully understood using a smaller unit of analysis, since the interdependencies between all of these factors are what transforms the system over time. CHAT sees human activities as *"complex systems under constant change and self-organisation"* (Engeström, 1990, p. 171; our emphasis), and defines grasping the essence of a thing, phenomenon, or person, as understanding how it has developed into what it is now (Vygotsky, 1978).

### 4.1 Engeström's Model of Learning by Expanding

Hedegaard et al. (1985) found that for some analyses of the transformation of activity over time, it is useful to specify the relations to "other subjects" in certain ways. Yrjö Engeström is one of today's most prominent users of CHAT for developing work in organisations; he uses the representation of a system shown in Figure 1. In this model, the "other subjects" factor is split into community, rules, and division of labour, and the "subject" factor usually refers to a collaborating group rather than a single individual (Engeström, 1987).

Engeström supplies this model to members of developing organisations to help them organise description, analysis and reflection upon their practice and guide the transformations of the system. For example, the model has been used in this way for health centres attempting to achieve a more useful way of providing their services (Engeström, 1993), and for cleaning work to become industrialised (Engeström & Engeström, 1986).

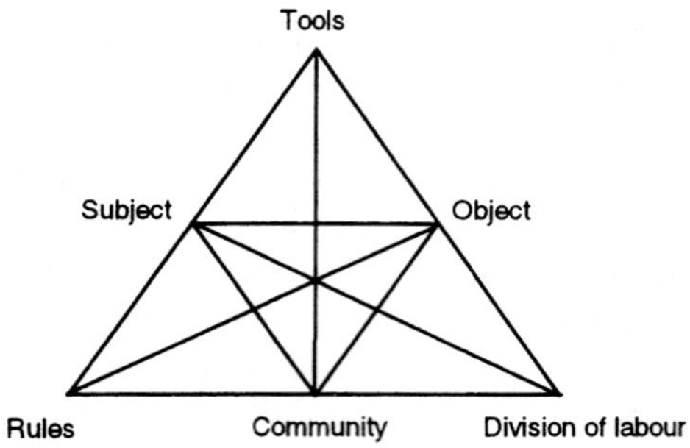


Figure 1: Engström's model of human activity system

According to the dialectical approach of Engeström, contradictions are what motivate and enable transformation. Thus, to develop the system, it is necessary to identify vital contradictions and find ways of transforming the system that resolves them. Engeström calls this *learning by expanding* (Engeström, 1987).

To simplify somewhat, a contradiction is a relation that, at the same time, encompasses an interdependency and a conflict. Contradictions occur within and between nodes of the system, and between the system itself and other systems that it is related to in vital ways. In principle, the relation between the system before and after a transformation is itself the fundamental contradiction, if one thinks dialectically: what the system becomes is both a function of how it used to be (dependency) and a rejection of how it used to be (conflict).

### *4.2 Zones of Proximal Development*

Vygotsky's work on cognitive development can inspire a useful way of thinking about knowledge and its development in a group. Vygotsky (1978) stressed that the level of a child's development within a domain is not adequately described by the level of complexity of tasks she/he can handle autonomously. In collaboration with adults or more knowledgeable peers, a child can participate in performance at a higher level of complexity than that which she/he can manage alone. The difference between autonomous and guided performance is called the zone of proximal development (ZPD), since, according to Vygotsky, "what a child can do with assistance today she will be able to do by herself tomorrow" (1978, p. 87). For a recent application of this view, describing how the process may proceed among adult learners, see Hutchins (1993). In brief, this perspective asserts that development of cognitive structure happens when the individual internalises a complexity that was formerly distributed over the system that she/he operates within.

Lewis (1995) suggests that this concept can be visually represented as shown in the upper line of Figure 2. The knowledge of an individual has a central core that is "owned" by the individual: it is available for autonomous use in the performance of tasks. Surrounding that core is the ZPD region, in which the individual has some knowledge but not the full structure of capacities required, and thus needs help in performing tasks that depend upon that knowledge. Core knowledge is that which is internalised already, whereas the ZPD represents the "system" (including other people and artefacts) in which the person participates at a level beyond her/his autonomous abilities in the domain.

Figure 2 also suggests that, when a community or group is considered, some parts of each person's core knowledge overlap with that of others and, most importantly, one person's ZPD may overlap with the core knowledge of others. From this model, one might conclude that the collective core knowledge is, not surprisingly, greater than that of an individual; but also that each person can support cognitive development in the group by providing "scaffolding" (cf. Wood, Bruner & Ross, 1976) for others.

This collective potential can only be realised if each member of the community is aware of the knowledge of others, and can capitalise on it, by offering and receiving

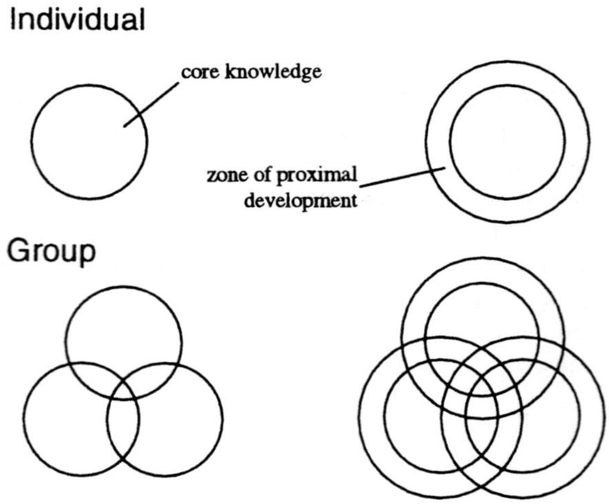


Figure 2: Core and ZPD knowledge for individual and group

help from the others. For effective working together, it must be appreciated that the full complexity of the knowledge of the group does not reside in any individual but is distributed amongst members of the group.

For CKC, this means that the individual participants—as well as the group as a whole—may benefit from working closely together. However, it is a prerequisite that the participants in the group have overlaps within their core knowledge that constitute a shared reference area or "recursive base" for exchanging information (Hansen, 1995). Also, they must continuously strive to sustain and enhance this overlap—a process that is sometimes called "grounding" (discussed in more detail in chapter 3, this volume). The CHAT model (Figure 1) further suggests that the group has to share objects in order to focus attention and direct the development; otherwise, the grounding process cannot work.

## 5 Relations and Contradictions

The previous section introduced the theoretical perspective of CHAT. We now discuss some of the issues and problems identified in the case studies, using this framework in the attempt to reach a deeper understanding.

### *5.1 Becoming a Group*

Engeström's model is a general description of the inter-related components in human activity systems. As such, the model provides a holistic picture of a collaborative knowledge construction process and its interdependencies, and can help to organise a thorough description of such systems.

Furthermore, the model is an aid to identifying potential vehicles of learning in a system: namely, the relations and contradictions that are inherent in it.

Space constraints do not allow us to present a full Engeströmian analysis at this point. By way of example, we therefore restrict ourselves to presenting fragments of such an ongoing analysis (Hansen, in preparation), as follows.

Health informatics is one of Aalborg University's recent part-time distance education courses, having started in 1994. The course differs from the one referred to in case 1, but, like the former, it employs problem-oriented project pedagogy and relies on a "classical" conferencing tool (in this case, the "first-class" system).

The theme of this course is information technology in the health sector, and students are encouraged to investigate problems that are related to their working life. In Denmark, a huge increase in the use of information technology within the health sector is expected, and a few hospitals are presently "project" areas with, for example, electronic patient file systems, which are not, however, fully implemented anywhere as yet. Aalborg University offers the health informatics course as an opportunity for health sector workers who are interested in these prospects, and who want to gain the knowledge they need for influencing the course of development in health provision activity.

Using Engeström's model, health informatics can be described in the following way. A new tool—informatics—is entering the system of health provision activity. As it is expected to become a pervasive part of the tool-node, it will change the nature of the system as a whole, i.e. it will transform health provision activity.

The health informatics course itself is a tool for health sector workers who are concerned with this change. In Engeström's terms, what motivates students to take the health informatics course is the contradiction between the present and the future state of health care activity. Taking the course is the first phase of their acting upon this contradiction: namely, qualifying themselves to participate in guiding the transformation that the contradiction will enable.

As one student put it, in a draft for the group's introduction to their project: "We are trying to find the right formulations of questions to the answer 'Information Technology'."

This particular project group consisted of two nurses and a doctor, who were working at three different hospitals. Their project concerned booking (assigning/scheduling patients for examinations and treatments), this being a process that will be transformed as a result of informatics systems becoming new tools for this task. The result will be a significant change in the practice that these students have known during their own working lives.

Specifying the nodes in Engeström's model with a few of the characteristics of the group, as shown in Figure 3,[3] immediately highlights some contradictions: i.e. some of the potential vehicles for these students learning by expanding.

3. For simplicity, in this figure we specify only the three nodes discussed in the text. Three-node excerpts like this can be used to emphasise a certain sub-set of the relations in a specific case (e.g. see Engeström, 1995). Lewis (1997) also uses triads of the nodes in a general description of sub-sets of issues arising in group learning contexts.

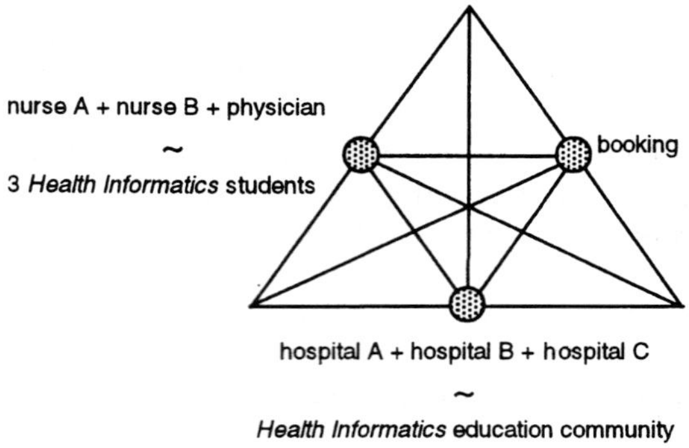


Figure 3: Four communities meeting in the booking group

At the subject node, the defining characteristics of the three persons with respect to the object "booking" poses the first contradiction: are they two nurses and one doctor, or are they three students? Closely interacting with this question is one concerning another node: what is the community? Is it the rest of the health informatics students and the community of Aalborg University, or is it the three respective hospitals in which the students still work, and from where they got their initial motivation to study booking?

The answer must be "both". Obviously, they are students in the health informatics community, for it is only within this community that this group is a group. But just as obviously, the motivation, knowledge, questions and concerns that these people bring to the group are grounded in their practices: i.e. within their respective work communities.

Thus, there is a within-node contradiction between two sets of essential features of the subject; and there is a within-node contradiction in the community node, which contains four communities, of which each person is related to two:

- the new, temporary one of health informatics, which is shared with the other two students in the group, and
- the more well-known and long-lasting one of their hospitals—this includes contradictions from differing practices of the persons when each is in their work community, performing according to the local rules and division of labour there.

What does this mean? Most importantly, it means that those involved have (partly) different professional concerns and perspectives and, from this everyday practice, bring different chunks of knowledge into the group with respect to the object "booking". Thus, at the outset, object-related knowledge chunks of potential relevance are distributed over the group. However, to make this functional, group members have to find common ground that can be used as a basis for informing and scaffolding each other with the privileged knowledge each possesses (cf. section 4.2).

Some of these characteristics also appear in the research group of the CSCW case study, which found that membership of a distributed community was not the only community membership of the individuals. Usually, the distributed community membership is supplementary to a more permanent, primary one. Furthermore, membership in the distributed community is temporary, often within a fixed time window. Indeed, a distributed community as a whole often has a predefined lifespan (for instance, 1 or 2 years). Amongst other things, this means that some of the conventions of the distributed community are *ad hoc* and pragmatic, to be invented as the work proceeds.

### *5.2 Reaching Understanding and Establishing Shared Goals*

The setting of shared goals is a well-known difficulty since this demands that each individual must assert their interests and concerns. This is where it is decided what the product and process of the collaboration should be and, thus, if concerns are not met then, they may never be met at all. Thus, it is in this phase that most emotional and motivational energy will be invested in making one's concerns explicit and negotiating to get them met.

There seem to be two main elements that need to join forces for a community to be able to work towards shared goals. One of these is predominantly a social one—a willingness to participate and share in a collective task; the second is basically cognitive—a willingness to share may be present but a fundamental difficulty may exist in developing a common language and understanding with the rest of the group.

The two types of CKC outlined differ in that, in the project pedagogy case, at the outset the students had to negotiate the very object of the collaboration (what exactly are we going to investigate?) as well as how to carry it out. In the other case, the researchers had a reasonably well-defined objective from the outset (coordinate their course module descriptions) and, apparently, only had to make decisions on how to carry it out. However, in both cases, when agreement was not already established, it is striking that what happened on the conferencing system was not what one would expect from people fighting over conflicting interests: participants apparently did *not* engage in a significant amount of negotiation in order to get their concerns met.

To ease description, in order to further analyse this paradox, we shall employ a model from the domain of pragmatics. Allwood et al.'s model of interaction claims that each of four levels of communicative functions are prerequisites for subsequent ones (Allwood, Nivre & Ahlsén, 1991; see also chapter 3, this volume). Simplified, the four levels are:

1. willingness to sustain contact,
2. ability to perceive message,
3. understand the message, and
4. agree on the content.

Some of the problems arising in both of the case studies seem to suggest either unwillingness to interact at all, or inability to perceive what the others said. For example, project learners each sent messages to the conference, but seemed to do so without taking into account what co-participants' messages said (see Henri,

1995, for a similar observation of lack of "real interaction"). Researchers sometimes seemed unaware that decisions had already been made and carried on as if they had not. In other words, at first glance, the problem appears to be at level 1 or 2 in Allwood's hierarchy. However, such an interpretation would be superficial. The participants were indeed motivated to interact, as shown by their organisation and participation in face-to-face meetings in order to sort things out. (For students, this was also confirmed by interviews; see Dirckinck-Holmfeld, 1990.)

Furthermore, by means of the arranged face-to-face meetings, the very same individuals could come all the way up the Allwood hierarchy and leave the face-to-face meeting with an agreement reached. This clearly demonstrates that there was no principal hindrance for reaching understanding and even agreement; i.e. it was not something inside these very individuals—or the combination of them—that blocked the process. The observed problems must have been a consequence of the communicative tools employed.

All issues requiring negotiation and decision making are difficult when participants originate from different communities, since, in such cases, it is not possible to understand fully the arguments of the others, due to diversity of background knowledge. In distributed or *ad hoc* groups, there is bound to be limited access to the daily practice of others, within which their knowledge and concerns are rooted. When the collaboration is mediated by telematic tools, the lack of "richness" makes clarification even more difficult—and, at the same time, provides a more shallow level of shared contextual resources to align interpretations.

Face-to-face communication and collaboration provides better opportunities for understanding. During copresence, there is continuous negotiation of meaning of utterances by verbal and non-verbal means, to "test" and confirm that the meanings remain aligned.

When the process of collaboration is mediated by telematic services, members of the group lose their shared context, and additional mental work is required in order to restore this. Sometimes, this may not even be possible if members are simultaneously affected by other activities in their separate communities. In any case, it may be very demanding, since it requires making explicit a lot of background knowledge that is taken for granted within face-to-face communication. In such cases, it must be very important to use communicative tools that do not require participants to spend too much of their attention on the tool itself: i.e. to use transparent tools (cf. section 2.2).

Moreover, in collaborative knowledge construction, some minimal structure is required to scaffold the alignment of meanings, and that structure needs to be sustained. This is much easier when feedback is immediate, continuous, and unambiguous. When the ambiguity grows, more information is needed to reduce it. Furthermore, it cannot be just any information, because some information may increase the ambiguity. Rather, it must be information that constrains the range of possible interpretations: i.e. reduces the ambiguity. More-sophisticated telematic tools attempt to accommodate this need by supplying shared (virtual) objects, such as "whiteboards" and "shared workspaces" (e.g. Aytes, 1995; Dix, 1997).

Star (1989) suggests the concept of a boundary object, which refers to those entities that "are plastic enough to be adaptable across multiple viewpoints, yet maintain continuity of identity" (Star, 1989, p. 37). That is, they must allow for local interpretations and concerns to be included, yet maintain a common core and shared constraints (cf. section 4.2). Such objects help and direct the evolution of ideas and keep the horizons of the participants adequately aligned.

We suggest that a good problem formulation can be seen as an important boundary object in CKC. To fulfil its role as a guide for the rest of the process, the problem itself must be made as concrete as possible. The problem must be specified in terms of a short description, a decision about which questions must be answered in order to constitute a satisfactory analysis, and an identification of theoretical (and sometimes empirical) material that can provide this result.

At universities using project pedagogy, problem formulation is a discipline in itself, and much effort is put into teaching students precisely that, since this is assumed to be one of the most helpful general skills for "knowledge work" regarding real-world open problems. However, to be able to construct a shared problem formulation, participants need a shared objective and mutual engagement in the task. The lack of real discussion in the problem-setting phase (case study 1) may indicate that it is too difficult to be achieved when using only asynchronous, text-based computer conferencing. However, an inadequate problem formulation often results in more frustration and delay later in the process, since the lack of an explicit common ground and focus to guide the work produces even more diversity in participants' perspectives, concerns and interpretations that will surface throughout the process and blur the focus.

In summary, to proceed as a group, participants must create a conceptual tool that represents a shared objective and possesses the characteristics of a boundary object. Using the "booking"-project group for illustration again, Figure 4 asserts that, in the first step of the CKC process, creating a problem formulation is in itself the objective. Once created, it becomes a tool for reaching the larger objective of understanding booking.

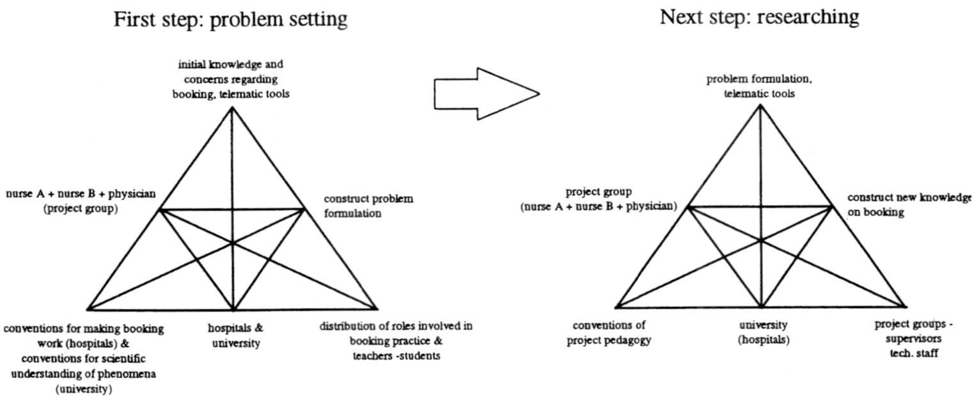


Figure 4: The first transition

This "first step" is not equally explicit in the CKC process of all groups. For example, for the researchers in case 2, the goals of the collaboration were formulated prior to the outset of the actual collaboration. Such *given* goals are, of course, external representations of a shared object. In this respect, they resemble the description of the problem formulation that a group arrives at when they create the formulation themselves. However, in the case of a given goal, participants cannot know to which extent their individual interpretations of this external representation are shared. The problem formulation *process* can be seen as a means to make several types of divergence surface at an early stage of the collaboration: divergent interpretations of the objective, divergent motives for participating, divergent preferences of method and standards, etc. Dealing with them at this stage reduces ambiguity later, and thus may be preferable also for groups who collaborate to reach given goals.

### *5.3 Needs Change over Time*

Although each process and product is in a way unique, most CKC projects (learning or working) proceed through a number of analytically distinct *phases*. These may be characterised as follows:

1. *Group formation*. The preliminary ideas are formulated and project groups are established.
2. *Problem setting*. The group carries out an analysis, through which the preliminary ideas are confronted with either the formal curriculum (teaching and learning) or the task to fulfil (working), and transformed into a systematic characterisation of the objective. This includes decisions about what specifically to investigate and what methods to use. This phase can be brief and relatively implicit if the group solves a given problem, or extensive and including much negotiation, as in the case of problem-formulation according to the project pedagogy method.
3. *Planning*. The investigation (research) is planned, and more or less division of labour within the group is established.
4. *Research phase*. This includes information searching, literature studies, and sometimes empirical investigations. The research phase is carried out according to the plan, but often in a process of confronting and managing unexpected obstacles, which can be related to any of the components of the system (the subjects, the tools, the object of investigation, and the community within which the activity is embedded).
5. *Conclusion of work*. On the basis of the research phase, the group draws conclusions in relation to the initially raised problem or goal. This often includes suggestions of reformulations of the problem or choice of another method that could provide more insight). The results of the project are written in a report.
6. *Evaluation*. The report is evaluated either through an examination (students) or through the community (work/peer review).

The overall direction is one of proceeding from phase 1 to 6. However, a closer look will reveal that interactive processes are embedded; e.g. after having carried out some research (phase 4), the group may have to revise the original plan (phase 3).

A central notion of activity theory is that human activity is organised at three embedded levels (Leontiev, 1978). In Leontiev's conceptualisation, the three levels coexist by definition (they are different levels of analysis for the same activity, i.e. analytical categories). This notion has been part of the inspiration for a range of researchers, some of whom interpret the three levels as representing a three-layered, hierarchical structure of separate events (e.g. von Cranach, 1982; Linard, 1995). Heeren and Lewis employ this way of interpreting the relationship between levels in order to suggest a temporal framework of CKC:

> "The *intentional* level is oriented toward motives: needs, desires, or values. It is the level of global orientation which gives meaning to human processes. An example from distributed research communities is the development of a shared understanding of research questions to meet a desire to gain new insights relative to a particular research topic.
> The *functional* level is oriented to specific, conscious goals in the context of motives. It is the level of focused organisational, planning and problem-solving processes in order to achieve a final goal or intermediate goals. An example from distributed research communities is the definition of the research methodology.
> The *operational* level is oriented to the practical conditions of actions. It is the level of practical routines which are a prerequisite to the conscious, purposeful actions at the functional level. An example from distributed research communities is the carrying out of an investigation according to the selected methodology." (Heeren & Lewis, 1997, p. 89)

The main assertion of Heeren and Lewis (1997) is that, over time, the focus of activity is at different levels. Figure 5 suggests a graphical illustration of this. Furthermore, these researchers suggest that ambiguity in a community is reduced as the task moves through the stages with intentional, functional, and operational focus, respectively. They use this assertion as an argument in favour of the view that "richness" of communication channels for effective collaborative working or learning needs to be high when the ambiguity is high (i.e. at the intentional level) and should be less rich when the focus moves to the operational level. This implies that different tools will be most efficient for different parts of the task.

The assertions of the two temporal models presented in this section—one based on phases, the other on levels—can be integrated in the view that different action levels dominate different phases of project work. As a consequence, the adequacy of a given tool will systematically vary with the phases of the CKC process. In phases dominated by functional- and operational-level action (e.g. the research phase), simple telematic tools are adequate, whereas phases dominated by action at the intentional level (e.g.

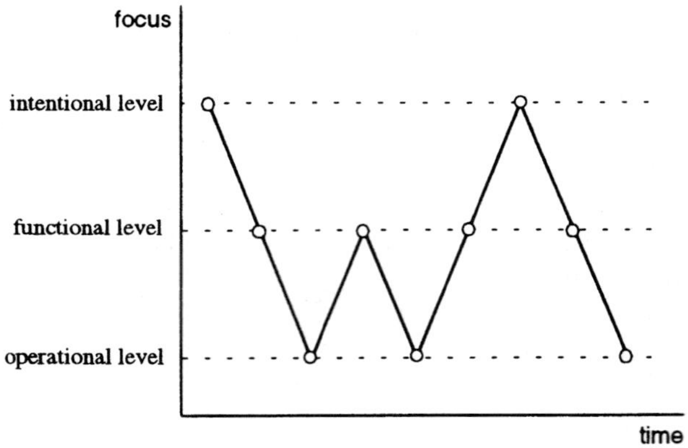


Figure 5: Over time, the focus of human activity is at different levels

the problem-setting phase) demand such a high degree of "richness" in communication that face-to-face meetings are required.

Further research on the integration of these models may lead to a taxonomy of telematic tools suited for different project phases. The property of recurrent cycles embedded in the process of project work (e.g. that plan revision may be required after having started the research) prevents a deterministic description, and asserts that *ad hoc* shifts of media should always remain an option. Nevertheless, such a taxonomy might be useful heuristically, allowing the planning phase to include a rough scheduling of different media usage over the project period.

## 6 Conclusion

The primary aim of this chapter has been to contribute to the understanding of collaborative knowledge construction (CKC). We believe that a deeper insight into the nature of CKC is necessary for understanding what roles telematic tools can play in the processes by which CKC operates.

Cultural–historical psychology/activity theory (CHAT) offers a theoretical basis for understanding CKC. This theory asserts that CKC is a complex, systemic phenomenon. To understand a CKC system, it is necessary to consider the people involved, the tools that they use, the objects of their thoughts and actions, and the communities that the activity is embedded in.

Using this framework, we have attempted to transform some findings from two case studies into more principled concepts within a theoretically framed discussion. In particular, we have discussed whether it is possible to identify different elements and processes in CKC that seem to require different kinds of tool support.

In our case studies, even simple, low-bandwidth telematic services could adequately support certain of the tasks involved in CKC. When participants carried out tasks according to previously set and agreed-upon goals, plans, and methods, the coordination of their efforts could be sustained by mainly asynchronous and text-based communication channels, such as "classical" conferencing tools. For such purposes, the available telematic tools compensated well for the problem of distribution of participants.

However, to actually set the goals and choose the means by which to pursue them, complex negotiation can be necessary. The classical telematic services available to the participants in our case studies seemed to be inadequate for this purpose. Attempts to negotiate such issues sometimes lacked real interaction between participants, surfacing as confusion about decisions, lack of attention to others' contributions, and inability to proceed until a face-to-face meeting had been held.

This led to the suggestion that different phases of CKC require different amounts of information "richness" in the communication. When a group must negotiate at the level of intentions, "rich" interaction is required to compensate for diversity of participants. However, when there is agreement at this level, more shallow channels of communication suffice. Thus, the need for "richness" will shift back and forth, but not in a random way.

In our case studies (as in most present-day examples of CSCL and CSCW), when "rich" interaction was required, face-to-face meetings were the only reliable means to obtain it. However, telematic services are presently being developed that allow much richer interaction than the classical conferencing tools that are presently in use in everyday settings. Unlike the "classical" tools, the future tools are transparent, and some of them are designed to actually facilitate human thought, besides compensating for obstacles, such as distance. These prospects are interesting, and future research and development should consider which levels and phases of the collaboration these tools can support.

CHAT sees conceptual (or "semiotic") tools, as well as material tools, as instruments that can facilitate human thought and collaboration. This indicates that the quality of the conceptual tools is as important as the quality of the telematic tools. We have suggested that problem formulations of the type advocated by project pedagogy are useful conceptual tools for CKC groups, because they require and sustain an operative agreement at the intentional level. We also suggest that in the continuous striving to develop better technical solutions, we should aim for deeper understanding of the processes that they are to support, in order to develop conceptual tools that can facilitate this work.

## Acknowledgements

This chapter was written as a result of the authors' participation in the European Science Foundation's research programme "Learning in Humans and Machines", for whose support we are grateful. In addition, while writing this paper, Tia Hansen received financial support from the Danish Council for Research in the Humanities under grant no. 9502778, and Jože Rugelj was partially supported by the Ministry of Science and Technology of the Republic of Slovenia under contract no. 3411–97-22–9085.

The "Virtual Mobility and Distributed Laboratories Project", which supplied some of the data used in this chapter, was supported by the Human Capital and Mobility Programme of the European Commission between 1994 and 1995. We would also like to thank the students and researchers who let us observe their work, Lancaster University for facilities made available to Jože Rugelj and Tia Hansen during their visits as Junior Scientists in the LHM-Programme, and Pierre Dillenbourg, Päivi Häkkinen, Kim Issroff and Daniel Schwartz for their comments on earlier versions of this work.

# 10

# The Productive Agency that Drives Collaborative Learning

Daniel L. Schwartz

In over 60 years of research, there have been very few demonstrations that working in a small collaborative group yields cognitive outcomes that cannot be matched or exceeded by the most competent member of the group (Hill, 1982; Hastie, 1983; Barron, 1992). This finding extends to tasks as diverse as brainstorming (Dunnette, Campbell & Jaastad, 1963), writing (Fox & Lorge, 1962), problem solving (Kelly & Thibaut, 1969), rope pulling, and rule induction (Laughlin & Futoran, 1985). The research suggests that there is nothing particularly special about working in small groups, at least with regards to cognitive outcomes like learning and intellectual artifacts. Yet, this conclusion seems to contradict common sense.

For example, I recently had a discussion with an ethologist, Dick Porter, who studies whether prenatal exposure to odor affects behavior after birth. He described studies in which chicken eggs are exposed to a particular scent (e.g. Turro, Porter & Picard, 1994). After hatching, groups of three chicks are placed in cages with a dish of scented grain and a dish of plain grain. Porter then observes their behavior toward the two dishes. What is of relevance here is why he uses three chicks per cage instead of one. Although chickens are cheap, they are not that cheap; and using three chickens as the unit of analysis increases the number of chickens he needs three-fold. He explained that the reason for using three chicks is that if you put a single chick in a cage it stands still. It loses all productive agency. It will not move towards either dish of grain in the attempt to learn the location of the preferred food. Chickens, like humans, are very social creatures.

In the case of the chickens, it is quite clear that two heads (or three heads) are better than one—better in ways that have profound effects on their abilities to learn. Yet, in the case of humans, there is little research that shows that 2+ heads are much better than one. What are we to conclude? That chickens are different from humans or that our small-group research has been missing something? Well, clearly the former is true, but so is the latter. This latter point is where the chapters in this volume come into play. As a collective, the chapters take on the very important and difficult task of trying to develop new research approaches that can help identify what is special about collaboration. The authors of the chapters draw on state-of-the-art research techniques involving robots and humans, and they tackle state-of-the-art problems, such as creating computer-supported collaborations. Bringing computers into the cooperative equation is a promising new approach to this area of study (see also Robertson, Zachary & Black, 1990).

## 1 Overview

In the following discussion of the chapters in this volume, I begin by describing what I think is especially fresh and exciting about the use of computers to understand cooperation. Next, I propose a picture of human nature that extends some of the ideas from the chapters and helps to counter-balance others. My picture will be distinctly psychological. I hope to show the fundamental importance of considering productive agency in our research on collaboration and collaborative learning. Just as the socially situated perspective has made it clear to cognitive psychologists that culture is not simply a background variable, I would like to suggest that research into collaboration should make it clear that agency is also not "simply a background variable."

One reason I have chosen to emphasize individuals when the topic is collaborative groups comes from an observation by Weiss and Dillenbourg (chapter 4, this volume). In their discussion of collaborative computer agents, they suggest that it may be difficult to prespecify all the rules or social conventions needed for collaborative behavior. Collaborative situations are often too complex. Consequently, the agents need to be able to self-improve and organize their collaborations. So, rather than exclusively describing the in-place conventions that regulate ongoing group behavior, one may inquire into the properties of individuals that make collaborative behaviors emerge. I believe the way to focus this inquiry starts with an understanding of productive agency.

My goal is to step back from the specific models of cooperation that have been generated by the authors in this volume and to explore some basic issues involving assumptions about the psychology of human nature. These issues—individual agency, opportunities to be productive, and constructive learning—were triggered and informed by the authors, but my discussion sometimes reorients the directions that the authors considered. One place where I would like to step back involves the authors' frequent wrestling with what counts as cooperative or collaborative learning. I do not wish to add to the clutter of definitions that Dillenbourg (chapter 1, this volume) heroically organizes in his introductory chapter. Missing, however, is the notion of agency that I believe is intimately intertwined with collaborations. Collaboration is not constituted by people following social laws as though they were physical laws. People need to choose whether and when to collaborate and whether to go beyond the minimum necessary to meet the rules of collaboration. I will attempt to argue that the very definition of collaboration involves the notions of agency and an individual's ability to represent other people's agency.

I also think that we need to consider the motivations that lead people to collaborate for the purposes of learning. There are many instrumental motivations, like joining to beat a common enemy, that are well documented in the literature (Slavin, 1983). But, I would like to focus on motivations that involve learning *per se*. Many of the current discussions view collaborative learning as the appropriation of ideas from others (implicitly, at least, this seems to be the motive). I wonder whether this is a satisfactory view of human collaboration. I want to propose another view—one that emphasizes the important role of being a contributor rather than a borrower. People appropriate knowledge when they are given opportunities to *produce* knowledge.

Finally, I want to explore issues of learning and collaboration, and the role of knowledge construction and language in this process. It is tempting to view some forms of collaborative learning as resulting from the direct communication of linguistic ideas from one agent to another. Computer models of collaboration, for example, often rely on the communication of propositions from one agent to another. But, one does not really learn a proposition, one learns *from* a proposition (Bransford & Nitsch, 1978). People use propositions to help them construct new knowledge. Linguistic representations can play only a partial role in learning from a partner. I would like to explore the nature of this partial role to help identify what knowledge is most likely to be generated in linguistically mediated collaboration.

All told, I want to paint a picture of humans in which productive agency plays a central role in the characterization of collaboration. One way this agency is expressed is by the decision to collaborate and the effort to make a collaboration work when social rules are not sufficient for successful collaboration. Another way this agency is expressed is by the motivation to produce and contribute. Finally, this productive agency appears in the very way we learn—we construct knowledge. In the following sections, I develop these ideas more fully and consider some of the ways that these characterizations may be addressed empirically. At the end of the chapter, I consider some of the ways these observations could be incorporated into computer models of collaboration and collaborative learning. But, first, I next suggest what is so special about computers in the context of understanding collaboration.

## 2 The Potential of Computers for Understanding Collaboration

The chapters herein offer excellent examples of how computers provide new energy for understanding collaboration. These examples include practical problems and potentials made possible by computer technologies like tele-communication (chapter 9, this volume) and robotics (chapter 5, this volume). These real technologies demand new methods of analysis (chapters 2 and 3, this volume). The eventual answers to the unique challenges of integrating computer technologies into human endeavors may have important, general implications for understanding collaboration.

A different example of how computers provide new direction to collaborative learning research comes from work in artificial intelligence and distributed artificial intelligence (chapters 4, 7 and 8, this volume). This work highlights the great potential of using the computer as a research tool in the tradition of cognitive science—as a way to model and elucidate the nature of cooperation among humans. In my discussion, I will focus on this possibility, because I believe the use of computers to model collaboration is a potent methodology. Regardless of one's theoretical disposition toward information processing, there are two important possibilities. Computer modeling, in general, may provide a way to handle the awkward size of small groups, and artificial intelligence techniques, in particular, may help to clarify the importance of representing the thoughts of others during collaboration.

### 2.1 The Awkward Size of Small Groups

Recall the earlier observation that it may be impossible to prespecify the group rules and social conventions that organize people during their various complex interactions and settings. While this problem may have something to do with human nature, it may also have something to do with the methodologically awkward size of small groups. This point may be understood in the context of a conversation I had with Joe Hamilton, a nuclear physicist at Vanderbilt. In relation to another project, I asked him why he thought mathematics has served modern science so well. His response was surprising. He said that in his domain, traditional mathematics was not working so well. He pointed out that we have excellent mathematical tools, like statistics, for handling large numbers of entities, as might be the case when talking about the behaviors of the many atoms and molecules in a gas. He also pointed out that we have excellent mathematical tools, like calculus, for handling small numbers of entities, as might be the case when predicting the behavior of a projectile in space. We do not, however, have a traditional mathematics for handling middle ranges of interacting entities, like the number of protons and neutrons in a nucleus. As a result, nuclear physicists rely on computational models more than other sub-disciplines in physics.

This seems to be most relevant to research involving small-group collaborations—collaborations that often involve several interacting participants (chapter 1, this volume). Unlike sociological data, small-group behaviors are not smoothed by averages over the thousands of people involved in a social movement. And unlike individual psychological analyses, the multiple interacting agents of a group cannot be turned into the simple equation of a single individual interacting with a single stimulus. It strikes me that computer models of cooperation may be powerful scientific tools that can help in the way that traditional mathematics has helped other physical and social sciences. They can provide a formalism that helps to define regularities and that can model complex interactions among medium numbers of entities.

### 2.2 Modeling the Thoughts of Others

The second reason that computers may make a powerful research ally, the reason that I will emphasize here, is that they permit a type of modeling that is ideally suited to characterizing individuals within collaborations. Specifically, artificial intelligence techniques allow us to model people's thoughts in the folk terms that they experience; terms like, "I agree," and, "This is my goal." This modeling ability may be extremely important to studies of collaboration—even more important than for studies of individuals. This is because a critical component of collaboration involves the representations that individuals have of one another's thoughts. As I argue next, this may be at the heart of understanding collaboration.

A common claim made throughout the chapters herein is that for collaboration to occur, it is necessary for the collaborators to have a model of one another's thoughts, and ideally for the collaborators to have a shared set of models. Weiss and

Dillenbourg speak of this necessity with regards to learning. They state "The 'deep secret' of collaborative learning seems to lie in the cognitive processes through which humans progressively build a shared understanding" (p. 75, this volume). If this secret is, in fact, critical to collaborative learning, then it seems that the capability of computers to model thought, even if poorly, is particularly relevant for understanding collaboration. This is because computers can model the way we construe other people's thoughts. In fact, this is exactly what computer models of cognition are; they are models of *other* people's thoughts. In this light, they can give us a chance to explore how models of other people's thoughts might affect collaborative activity.

Of course, learning through collaboration does not always require the representation of the mental states of one's partner. As Joiner, Issroff and Demiris (chapter 5, this volume) point out, there are different types of joint activity and learning. In some situations, I do not need to know your thoughts, I only need to know the consequences of your actions. For example, in a foraging task, robot A can coordinate with robot B by simply noting where robot A is and has been. More generally, there are many situations, like traffic lights, where people coordinate and learn behaviors by relying on social conventions. But, as I will try to explain later, these types of joint activities are not really what we have in mind when we think about collaboration.

Moreover, it is clear that we do learn to use other people's mental states to further our collaborative endeavors. Krauss and Glucksberg (1969), for example, developed a screen referenced task in which participants talking from either side of an opaque screen try to put their respective sets of nonsense shapes into identical orders. Although each partner has the same set of shapes, the shapes are in different orders and do not have conventional names. Therefore, the participants need to infer and represent the shape that their partner has in mind in order to complete the task. Krauss and Glucksberg found that there is steady improvement in collaborative performance with age: 5-year old children show little success, 6-year old children show some gains, 8-year old children show more success, and so forth. So, even if modeling another person's mental states is not a necessary prerequisite for learning in groups, it is clear that it is something upon which humans rely. The computer, with its ability to model the way that humans represent other people's thoughts seems ideally suited to sharpening our thinking about collaboration and collaborative learning. At the end of this chapter, I suggest some possible explorations. But, first, it is important to decide what types of traits a computer agent would need in order to model human collaboration.

## 3 Agency: From Compliance to Choice

To begin development of the notion of productive agency, I would like to argue that the very notion of collaboration depends on individual agency or intention. Many of the authors in this volume are concerned with a definition of "collaboration" and realize that it is important to begin to understand what does and does not fit this category. In this section, I provide some contrasting cases in an effort to tease out some of the features that seem important to understanding what is essential about collaboration.

One contrast comes from Mephu Nguifo, Baker and Dillenbourg (chapter 7, this volume) who raise the issue of global versus local control in a cooperative interaction. Computationally, we might ask whether it is better for robot agents to follow globally specified procedures, or is it better to let cooperative behaviors emerge from procedures local to the individuals. In human terms, one might think of this as the issue of whether individuals or social conventions dominate collaborative interactions. Naturally, it is both, and the balance should change depending on the situation. Regardless, I think the essence of collaboration revolves around local control.

There are two types of local control. Local automata, like termites, have simple rules that, when coupled with other termites, can lead to complex structures. The global structure of this behavior arises from locally determined rules. The computer models used by the nuclear physicist might fit this category because the local properties of protons, neutrons, and so forth determine the global patterns they make when interacting. This situation is different from situations involving agents with local autonomy who choose to take on a particular behavior. It is this latter notion of autonomy that is particularly important for collaboration. If I am bound to a social role or predetermined sets of local rules, then it is difficult to say I am collaborative; instead, I am complying without agency, something like termites in a nest. To collaborate, individuals have to enter into relationships, they have to produce ideas, they have to choose whether to communicate, and they have to choose whether to compromise their goals.

In the next sections, I will develop the idea that collaborative learning is not constituted by people simply complying to a role. Collaborative learning takes agency and productive effort precisely because people must develop shared meaning across the differences in their roles and knowledge. To further this point, I will argue that the very definition of collaboration involves the idea of intent and non-compliance. First, I begin with the notion of "the effort after shared meaning," then I move to the idea of non-compliance. Hopefully, this discussion can help anchor the slippery notions of collaboration and collaborating agents. For example, it may provide a solution to the question of whether we should consider the neurons in the brain as collaborative.

### 3.1 Identity versus the Effort after Shared Meaning

A second contrast I would like to draw distinguishes between what is unique about research on collaboration relative to other social topics. In many different social–cultural traditions, a common question is how an individual fits into his or her cultural milieu through structures like roles, rules, and social practices. In terms of psychological issues, a key construct here is the notion of identity. An individual gains his or her identity by taking on a role offered by a culture or social practice. Identity is a bridge between the local individual and the global cultural milieu. It is the desire to gain an identity within a cultural milieu that causes an individual to appropriate and come into compliance with the practices and roles of a culture.

While identity and compliance with cultural practices are surely central aspects of humanity, there are others. To borrow Bartlett's (1932) phrase, "the effort after meaning" is also important. Piaget, as well as most of cognitive science, investigated people's pursuit of meaning and its effect on learning. For example, when people understand the meaning of a text passage, they remember more (Bransford & Johnson, 1972). The importance of people's effort after meaning also appears in social settings where there is "the effort after *shared* meaning." When I talk to my family, friends, or colleagues, I want them to understand me, and I want to understand them. The desire to understand and be understood—to share meaning—is a strong motivator of human behavior and worthy of the status of a basic psychological construct. In terms of cognitive science, one might say that individuals want accurate representations of other people's thoughts, and individuals want other people to have accurate representations of their own thoughts. Baker et al. (chapter 3, this volume) point out that we can never actually reach this idealized shared meaning in any absolute sense. But, they explain that it is the effort after shared meaning that helps explain why we learn when we collaborate.

Many of the questions for research on collaboration should be about how individuals construct local interactions among themselves to understand one another. Such questions are about how individuals interact with individuals, not how individuals interact with "culture." This is not to say that we should ignore the milieu in which these interactions take place, or the social rules that make exchange possible. Nor is this to say that desires for identity within a group do not play a strong role in group behaviors (e.g. the desire for the status of being team captain). But, there are times when it is worthwhile to minimize the emphasis on the cultural environment in which we swim, and over-emphasize people's efforts for achieving shared meaning. Hansen et al. (chapter 9, this volume), for example, describe a situation in which people with very diverse and distant "base communities" temporarily come together electronically to work on developing joint projects. The individuals do not have a solid common culture to regulate their interactions. Conceivably, the community differences serve as important forces that cause people to learn through collaboration. Several authors in this volume, for example, propose that misunderstanding *is* the progenitor of collaborative learning. When people cannot rely on common ground, they need to make the effort after shared meaning, and this often entails learning about and negotiating understanding with one another. For example, individuals in one base community need to learn the merit behind another community's goals.

Collaborative learning often occurs exactly when people cannot assume the common ground of similar thoughts, roles, and goals. Conversely, collaborative learning often fails to occur when people overly rely on social roles. A nice example of this was recently brought to light by Kathleen Jones who studies parent–teacher interactions (e.g. Hoover-Dempsey & Jones, 1996). She explained that in parent–teacher meetings at school, parents and teachers often remain in their roles and maintain very different goals for the child (e.g. to make it through the year without further disruption versus to support a happy childhood and nurture a healthy adult). Because of the roles that the teachers and parents maintain in these meetings, they often do not attempt to develop a shared understanding that can help them learn about the child's needs and behaviors from one another. The parent, for example, may passively listen to the teacher's evaluation and prescription. Clearly, an important question broached by nearly all the authors in this volume is when and how different types of affordances, like discourse pragmatics and visual representations, can induce people to make the effort after shared meaning.

One of the theoretical problems of exclusively emphasizing the social conventions and rules that support collaboration is that we can slip into the study of local automata without autonomy. In such theories of collaboration, we do not factor in the agency involved in collaboration. For example, think of the slaves who built the great pyramids. One would say they complied with their masters' rules and the prevailing social conventions (on pain of death), but one would not say they collaborated. Of course, the slaves may have collaborated with one another; e.g. by helping to carry a load when they did not have to. But, simply following a social role does not make behavior collaborative, it simply makes it compliant.

We can carry this point further by contrasting the "effort after shared meaning" with "having a shared meaning." Consider a definition of collaborative activity that many of the current authors adopted and that is superficially consonant with the notion of the effort after shared meaning:

> "[Collaboration is] a coordinated, synchronous activity that is the result of a continued attempt to construct and maintain a shared conception of a problem." (Roschelle & Teasley, 1995, p. 70)

This description strikes me as problematic. It does not capture the essence of collaboration as it appears to the individuals involved; it provides no room for the agency with which people choose to collaborate or not, or choose to make the effort to understand one another or not. It does not capture the sense of compromise and choice that is the hallmark of any collaboration. By this definition, for example, we might have to say that the slaves were collaborating with their bosses as they built the pyramid. After all, they did maintain some shared understanding of the problem—to build a pyramid.

### *3.2 A Turing Test for Collaboration*

I would like to develop another argument for why we may want to factor in an individual's agency in our models of collaboration. I begin by pointing to a second problem with the definition of collaborative activity given in Roschelle and Teasley. In that definition, collaboration is primarily defined by its outcome, "coordinated, synchronous activity." It is not clear to me what constitutes coordinated activity. People have their own reasons for collaboration and their own definitions of coordination, and these reasons and definitions often change during the course of an interaction. Unless the researcher is willing to impose particular social norms as to what constitutes a successful collaboration, it seems that a general definition of collaboration should be defined with an eye toward the view of those people involved rather than the view of the prevailing culture as identified by the researcher. As I argue next, taking the viewpoint of the people involved leads to a definition of collaboration that is grounded in the notion of individual agency and intent.

If we view collaboration as something that involves individuals representing one another's thoughts, then perhaps the most relevant definition of collaboration would be from the eye of the beholder, or representer as it were. So, how would I, as the beholder, decide whether I was interacting with a collaborator? One approach to this question is to consider the traditional artificial intelligence Turing test to see if we can develop a "collaborative Turing test." The original Turing test was designed to answer the question of how we can decide whether a computer is intelligent. The test stated that if we cannot discriminate between a computer and a person (e.g. when interacting over a tele-type), then the computer should be considered intelligent. Weizenbaum (1966) made fun of this test with a computer program called ELIZA. ELIZA faked being a therapist. It had a number of generic sentence frames that it would use in response to anything a "patient" typed. For example, if the patient typed "I'm very unhappy about my mother," ELIZA would respond "*How long have you been* 'unhappy about your mother'?" It could just as easily say "*Tell me more about* 'your mother'." People were fooled by ELIZA and thought a human was on the other side of the tele-type. Weizenbaum concluded that the Turing test was absurd.

Since the original demonstration, there is now a software program based on ELIZA that is doing quite well in the stores. Even though people know that ELIZA is not a person (and not "really" smart), they still choose to use it. Evidently, people are willing to view it as a "smart tool" (Norman, 1993), even if they do not view it as smart. What is the difference between these two—being a smart tool and being smart? One answer is that people do not view a smart tool as a smart agent that is trying to fulfill its own goals; they simply view it as a clever device.

Perhaps the Turing test should not be whether people can discriminate between computers and humans; one can always manipulate people's short-term success at this task anyway. Rather, the test should be whether people are willing to interact with something as an intelligent agent rather than a "smart tool." To borrow from Dennett (1987), maybe the criterion of intelligence is whether people take an "intentional stance" toward the computer. In their interactions, do they "endow" the computer with an intelligence that they believe the computer uses to fulfill its intentions?

If we map this idea into a collaborative Turing test, then a piece of the criterion for identifying a collaborative partner is whether people are willing to view their partner as a collaborative *agent* during their interactions with it. Do they, for example, assume the computer has collaborative goals that it intends to fill?

People's intentional stance is an important component of collaboration; people need to perceive their partner as a collaborative agent rather than a supportive tool that simply follows or embodies a set of behavioral rules. But still, this definition of collaboration seems empirically toothless as it stands. It provides little insight into collaboration except that it involves the perception of intentional agents with goals. So, I would like to try a slightly different approach. Again, I begin with the question of intelligence. By some accounts (Eco, 1984), symbolic intelligence is characterized by the ability to lie. Consider work with chimpanzees. Researchers release one chimp into a field where there is a stash of bananas. After finding the bananas, the chimp returns to the band and points to the location of the food for the other chimps to share. So far, there is nothing particularly "intelligent" about this behavior; bees do the same sort of thing. One time, however, legend has it that a chimp came back and pointed in the opposite direction of the food. While the other chimps followed his misdirection, he went back to his now personal stash of food. This misdirection lie depended on the original chimp knowing something about the "thoughts" of the other chimps; namely, that they would follow his pointing. This seems distinctly intelligent. The relationship between intelligence and lying makes a good bridge into issues of collaboration and the effort after shared meaning. This is because intentionally lying depends on the ability to construct a representation of another person's (or chimp's) thoughts.

Now, let us combine the observation about lying with the importance of the intentional stance to come up with a collaborative Turing test: a partner is collaborative if you believe it is possible that the partner could be non-collaborative. In other words, you believe the partner has enough knowledge of your mental states and enough personal agency that it could intentionally thwart you or choose to disengage.

Even though this definition of collaboration is circular and surely not sufficient, I think it does tell us something. Collaborative behaviors probably depend a great deal on the extent to which people trust one another's cooperativeness. Will you tell your deep secrets to your therapist, "Eliza," if you think she will tell other people against your wishes? Will you be willing to collaborate with another scholar before you have had a chance to assess their intentions and beliefs about intellectual property rights and the ownership of joint products? If collaboration depends on representing another person's mental state, then surely one of the mental states we appraise is the other person's goals and likelihood of collaborating at any moment.

Consider the problem of remote collaborations recounted earlier (chapter 9, this volume). People from different communities need to cooperate electronically on a new project. The authors state that at the beginning of this cooperation, it is important to have a high bandwidth for exchanging information about constructing joint goals and administrative procedures. An interesting question is whether some of the difficulty people have in early telematic collaborations comes from their difficulty in developing mutual trust of one another's intentions. I wonder whether early interventions that led to trust among the participants could alleviate some of the early need for a high bandwidth of information flow. It is amazing how effective a relaxed dinner conversation can be for facilitating future collaborations.

The collaborative Turing test also provides some useful analytic distinctions among ideas that frequently become conflated. First, it points out that communication and collaboration are distinct. Imagine that you are in perfect communication with your word-processing machine. The computer records every keystroke perfectly. Does that mean the computer is collaborating? Not really, unless you believe it could intentionally put the wrong letter on the screen every now and then. Or, take the converse, imagine that I speak Hindu and you speak Japanese. Even though we would have miserable communications, this does not mean we are not collaborating. I know, for example, that if you get frustrated enough, you might choose to quit. Although communication and the effectiveness of a collaboration are empirically correlated, they are still different things.

Second, the definition helps to clarify that outcomes are not sufficient criteria for identifying collaboration. If basketball team A continually loses to basketball team B, this does not necessarily mean that team A is less collaborative. In fact, one might suspect that team A needs to be much more attentive to collaborative issues because there is a high risk of people becoming uncooperative in a failing group. Finally, I think the collaborative Turing test helps to eliminate some of the slippery slope that occurs when we consider embedded systems, like individuals in a group, termites in a nest, and neurons in a brain. I see no sense in saying that the termites or neurons are collaborating. They have no choice in what they do.

## 4 Motivation: From Appropriate to Production

The ability to express agency plays an important role in people's motivation for, and benefit from, collaborative learning. First, I think people need to have the intent to learn while interacting in a collaborative group; learning is not automatic (e.g. Bereiter & Scardamalia, 1989). Although there is a place for incidental learning outside of one's intent, I suspect that in many cases it is not enough for people simply to collaborate; they also need to have the intent to learn about the situation over which they are collaborating. There are many examples in which people have scripted cooperation among children, and the children end up producing the script rather than generating the knowledge they are supposed to learn through cooperation (Vye et al., 1998). Second, and more profoundly, I think people are motivated to collaborate to the degree they are also able to exert their agency through productive behavior. First, I will discuss the intent to learn, then I will take up the issue of production and motivation.

### *4.1 The Intent to Learn*

A wonderful example of the importance of the intent to learn comes from two different stories involving Eskimos. The first story involves the question of why an ancient European settlement in Iceland disappeared. One hypothesis is that, for whatever reason, the Europeans did not appropriate the knowledge of the Icelandic Eskimos and, consequently, the Europeans did not survive. This shows that the failure to appropriate can be a big mistake. The second story involves Eskimos and Athabascan Indians in Alaska. In a location called the Middle Yukon, the Indians are separated from the Eskimos by a thin mountain range. In times past, the Indians purportedly crossed this range and sneaked into the Eskimo village to kidnap women (McClellan, 1971). They did not appropriate the Eskimo women because there were not enough Indian women. Rather, they took the women because the Eskimos had developed excellent technologies. The Indians appropriated the women to gain access to their technological know-how. Interestingly, the Indians had to do this often. When a woman died, they would kidnap another. This is because they never bothered to learn what the women knew; they simply appropriated the technology and not the understanding.

I like these examples because they highlight that there is a difference between a culture of appropriation and a culture of learning. I think it complements the observation made by Baker et al. that learning is more likely to occur to the extent that agents expend greater cognitive effort towards mutual understanding than that which would be minimally required for communication. Evidently, the Indians were happy to communicate with their Eskimo women and did not put in the intentional effort needed to learn.

### 4.2 Putting Production Back into the Dialectic with Appropriation

What is it that causes people to put forth that extra cognitive effort toward mutual understanding? There are many things ranging from potential rewards to the desire for a friendly chat to the environmental affordances that support communication. In terms of basic motivations, I would like to illuminate the importance of production and original contribution. This appears to have been ignored lately, perhaps because of our over-emphasis on the appropriation of cultural practices. Appropriation is clearly an important idea, but it is only half of the story. Marx (1939/1973) spoke of two great forces that help constitute a person. One was appropriation—we become what we are by appropriating the ideas and artifacts of those around us. Alienation, his construct of psycho-social malaise, can occur when we are not allowed to appropriate the contributions of others. But, Marx did not consider appropriation to be the "key" quality of what it means to be human. Instead, he felt that humans are quintessentially builders. We want to produce and create ourselves in the world through our ideas and our material products. This way, we may put our "element" in the social matrix, and other people may appropriate our ideas. At the same time, we may "re-appropriate" our creations as they have been realized in the world, culture, and others. This serves as feedback about ourselves, our learning, and our environment. Without production, there is no feedback.

For Marx, the key to a complete person was not simply access to the material and intellectual wealth of a society. It also included access to the means of production. Marx did not advocate a welfare state in which people only had opportunities to appropriate. He advocated a productive state where people could contribute and impress themselves upon the world. Individuals are builders of their society, not simply recipients. For Marx, the key issue was who had the means of production.

The emphasis on individual production is an aspect of Marx that appears to have faded somewhat in the communitarian tradition that comes through Vygotsky and Engles. In Vygotsky, we read about the movement from external to internal, but we less often hear of the movement from internal to external. Although communitarian scholars speak of the importance of "activity," it still seems that the individual's productivity falls into the background. The emphasis on appropriation at the expense of production strikes me as particularly problematic when we consider collaborative learning.

Consider the World Wide Web. What is it that makes for a successful website? One thing is the opportunity for people to contribute. Hagel and Armstrong (1997) argue that successful websites are those that enable virtual communities. Whereas Prodigy emphasizes news broadcasting and information delivery, virtual communities support contributions by those individuals who use the website. Like the old adage, "The best conversationalist is a great listener," the most inviting website is the one where people can contribute. *Amazon.com* is a very successful on-line bookstore—in part, because it allows individuals to submit their own reviews of books. People contribute and they look forward to the responses. This is one way that we come to learn, by creating and re-appropriating the feedback from our creation.

As another example, consider what makes you most content in a conversation. Is it when you have been told something and understand it; when you have appropriated someone else's idea? Or, is it when you have contributed substantially to the conversation, when you have produced ideas that move other people and that help contribute to the direction of the interaction? Stated less rhetorically, perhaps the most irritatingly uncooperative agent is really the one who denies you agency within the group.

I think there are some wonderful research opportunities here. For example, one might use a simple version of Csikszentmihalyi's (1990) method of measuring flow. Every now and then, interrupt people during a collaborative episode and ask them to rate their "intellectual energy for learning" on a scale from one to ten (Schwartz & Bransford, 1996). I predict that, on average, people will give much higher ratings when they are contributing their ideas to a receptive group than when they are listening to somebody else's ideas. Moreover, I suspect that when people feel they are contributors, they will be much more willing to go beyond "the minimum necessary" to communicate and complete a task.

The dialectic between production and appropriation suggests that the opportunity to produce should influence people's willingness to appropriate from those with whom they are collaborating. One nice example comes from a study by Kay Burgess (personal communication). Burgess worked with fifth-grade students who learned how to solve a complex problem about rescuing an eagle. Later, those students came in as consultants to help college students solve the same problem. The fifth-grade students were empowered and highly engaged as they explained some of the intricacies to the college students. And, importantly, they became aware of and started to appropriate the college students' behaviors, like their diligence. Similarly, my colleagues at the Learning Technology Center at Vanderbilt have found that teachers become more involved in learning about new instructional techniques after they have had an opportunity to present their versions of those techniques to a new group of teachers. A study that formally examined the dialectical benefits of production on subsequent appropriation would make a profound contribution to the literature on collaborative learning.

## 5 Learning: From Assimilation to Construction

One of the reasons that production is so important to collaborative learning is that learning itself is productive. People construct their knowledge through generative mental and physical activities. They do not simply assimilate someone else's knowledge or practices; they actively produce their understanding. The constructive nature of learning has implications for how people learn, how they come to understand one another, and what they are likely to learn in groups. In particular, there are a set of implications surrounding language that are especially relevant to collaborative learning because collaborative learning typically involves linguistically mediated communication.

### *5.1 Some Background on the Relationship between Language and Learning*

As a starting point for understanding the implications of linguistic communication, we can begin with the chapter by Mephu Nguifo, Baker and Dillenbourg (chapter 7, this volume). These authors explicitly compare machine learning operators and dialog operators. They point out that there is a similarity between the two because both require the assumption of a common language. By common language, they mean that there is no ambiguity in symbol-referent mapping. If you are told "John is a cow," and you know the referents of the terms "John" and "cow," then you can assimilate "John is a cow" into your knowledge base. The work by Hoppe and Ploetzner (chapter 8, this volume) helps to clarify the point further. They describe a computer simulation in which one agent has a qualitative representation of a physics problem and another agent has a quantitative representation. Although the two computer models have different conceptual representations, they are able to communicate because their references to the problem at hand are unambiguous.

Weiss and Dillenbourg (chapter 4, this volume) point out that systems of distributed artificial intelligence depend on perfect symbol-referent mapping. They then argue that the necessary absence of referential ambiguity in these systems makes them incapable of modeling an important dynamic of collaborative learning; the systems cannot negotiate meaning. In other words, when the assumption of a common language (perfect referent-symbol mapping) is violated, computers cannot recover and learn from one another in the process. On the one hand, this appears to be a limitation of symbolic computer models; they require a common language to collaborate and exchange information. On the other hand, this also appears to be a strong limitation in humans as well. Baker et al. (chapter 3, this volume) point out that when people cannot assume that their expressions are understood by their partners, there needs to be an effort dedicated to reconstructing a basis for mutual knowledge. Often times, this effort falls short, just like in computers. But, as both sets of authors point out, it is in the attempt to rectify misunderstandings where a great deal of learning can occur.

Language plays an important role in the attempt to establish understanding among partners. To explore what role language plays, particularly with respect to learning, consider the discussion started by Ploetzner et al. (chapter 6, this volume). In their chapter, they compare the effects of self-explanations with the effects of receiving and giving explanations to others. By looking at this clean contrast, they hoped to identify essential differences between working alone and together. They do not find many differences, although they smartly speculate that feedback should be a key difference between self-explaining and other explaining (despite the fact that studies have not investigated this possibility). In addition to the lack of feedback, a likely explanation of these null results may have to do with the heavy use of language in both settings. Language and linguistic representations play a particular role in generating new knowledge and may be a primary mediator of learning effects, whether alone or in collaboration. To develop the point, I first explore the sorts of things language can communicate when people try to learn with or about one another.

### *5.2 The Implications of Constructivism for Reaching Shared Meaning*

When someone says something to me, I do not simply assimilate or copy that expression into my mental network. People do not learn a text, they learn *from* a text (Bransford & Nitsch, 1978). The words that you express serve as a starting point for me to construct or generate my own knowledge. Even if we come from the same culture, there is a distinct possibility that I will come up with something very different from what you had in mind when you uttered your expression. Consider, for example, the following newspaper headlines that were collected over the past year:

- Drunk Gets Nine Months in Violin Case
- Survivor of Siamese Twins Joins Parents
- Iraqi Head Seeks Arms
- New Study of Obesity Looks for Larger Test Group
- Kids Make Nutritious Snacks
- Miners Refuse to Work after Death

I doubt the newspaper editors had the amusing alternatives in mind. The sentences show that people do not simply assimilate the language of others. They actively generate meanings—meanings that may be quite different from what the speaker had in mind.

Although language is notorious in this regard, it is important to remember that perceived objects suffer the same fate. They are not simply assimilated into the head like photographs. People actively generate understanding using the input of the physical world just like they do from the verbal world. If you and I look at the same thing, there is no guarantee that we will see the same thing. The mere presence of the physical world does not ensure a common ground between two people who both have access to that world.

Consider, for example, the squiggles in Figure 1. Imagine that your task is to memorize them for an upcoming recognition test. Also, imagine that the artist who drew the squiggles is at your side. The artist would probably see something very different than you do. Fortunately, breakdowns in shared meaning are not

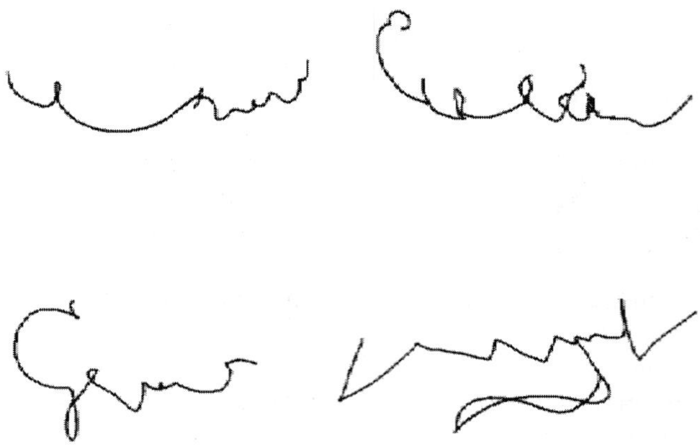

Figure 1: Do you see the same thing as the original artist? (Adapted from Gibson, 1969)

irreparable. In the current case, to improve your memory, the artist could probably help you to see what he sees with a little bit of supportive language. For example, "Turn the figure clockwise 90° and match each of the following labels to its respective squiggle: James Dean, Babyface, St. Nick, Baseball Bob." Let me reiterate this point more tele-graphically. Hoppe and Ploetzner (chapter 8, this volume) describe a system that uses a joint blackboard as a source of common ground for conversants. They describe the system as WYSIWIS—What You See Is What I See. Although not as easily said, I think the system is more appropriately called WYASIWIAS—What You Are Shown Is What I Am Shown.

Understanding is generated and constructed. The implication of this is that physical reference suffers the same fate as words; neither guarantees common meaning and learning between individuals. Even so, despite this similarity, there are very real differences between words and objects. Language helps in constructing a particular kind of knowledge. Language and other symbolic representations, like mathematics, are very good at helping people to build and evaluate an articulable structure. Moreover, there are criteria of understanding that come with symbolic explanations. Ploetzner et al. state, "Explanation is a social criterion for operationalizing what is accepted as understanding" (p. 104, this volume). I would like to amend this just a little to state "what is accepted as *structured* understanding." Most people, for example, have a very good understanding of how to ride a bicycle although they cannot explain it.

To demonstrate the power of language with regards to structure, consider Figure 2a. What do you see? People see many different things including a plane flying sideways and two olives on a toothpick. But let us assume that it is two men riding a tandem bicycle while wearing large, round hats. You are looking down at them. Next, consider Figure 2b. What do you see? If you are like most people, you see two pairs of men riding bicycles. Now, notice how language helps to restructure your thought. It is

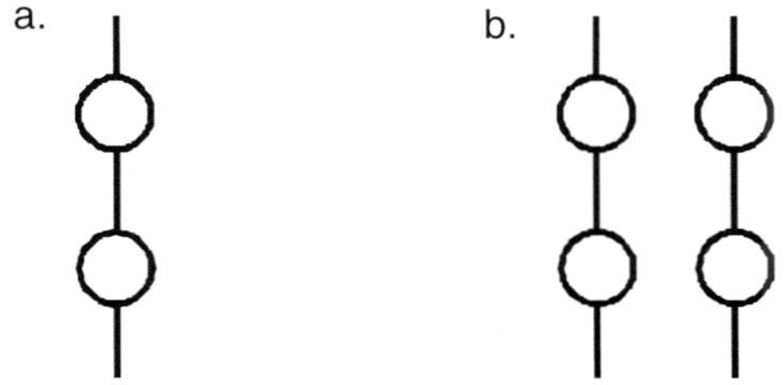


Figure 2: What do you see in the two figures?

not really men on bicycles; it is a bear cub clinging to the back side of a tree. The circles are its paws. Suddenly, new structures in the referent become important, like the distance between the two lines that portray the tree.

### *5.3 The Type of Knowledge People Often Generate in Collaborative Learning*

Collaboration typically involves heavy doses of language. This language should lead cooperative groups toward structural descriptions in their learning. Moreover, groups may move toward abstractions as the members try to find a safe place to communicate where their idiosyncratic differences of interpretation will not get in the way. It is this pull toward abstraction and structure in the verbal communication of groups that strikes me as the sort of place to find a special effect of collaboration on cognitive outcomes.

Consider the following study (Schwartz, 1995). Seventh-grade students were given several descriptions of fictitious fish and their habitat requirements. For example, "the spotted frolling lives in lakes with weeds," "the one-finned halluck needs weeds and a sandy bottom." The students' task was to construct a visualization of the various relationships. Students worked alone or in pairs. The students who worked alone drew pictures of lakes with fish in them. Only 6% created visualizations that were abstract in the sense that they did not actually look like fish and lakes. In contrast, 67% of the pairs constructed an abstract representation like a matrix or chart. This percentage is well above the probability that a pair would have included at least one member who would have constructed an abstract representation when working alone. In other words, the collaboration among the pair members led them to generate something new that was not found in otherwise similar individuals.

Another study, completed for this chapter with Doug Holton, demonstrates the point again, only in a negative fashion. The basic task involves glasses filled with imaginary water. Imagine that there are two glasses of the same height that are filled to equivalent levels of imaginary water. The only difference is that one glass is thin and one is wide. Would they start pouring at the same or different angles? Figure 3 provides a two-dimensional representation of the problem. In prior research, Schwartz and Black (1999) found that only 20% of individuals make the correct explicit judgment that the thin glass needs to be tilted farther than the wide one.

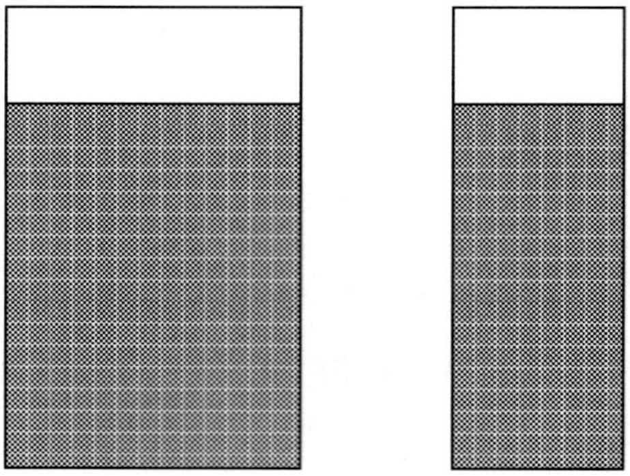

Figure 3: The shaded area represents water in the two glasses. If they are tilted, do they start pouring at the same or different angles?

When solving the problem explicitly, many people abstractly compare quantities like glass width, and this abstract approach leads them to faulty conclusions. However, when otherwise identical people close their eyes and tilt each empty glass in turn until they "see" the imagined water reach the rim, 100% correctly tilt a narrow glass farther than a wide one. The question addressed by the following study was what would happen when people worked together to solve the problem? One possibility is that one member of the pair would tilt each glass, and the other member would observe the different angles of tilt. Another possibility is that, because of the pull toward abstraction in group communications, the pairs would rely on more discrete and explicit reasoning and would therefore get the problem wrong.

Twelve pairs tried to solve the task. They were given the wide and thin glasses and were asked to figure out whether the glasses would start pouring (imaginary) water at the same or different angles. Interestingly, none of the pairs answered the problem correctly. The pairs did not imagine the behavior of the water, which is how people usually reach the correct answer. Instead, they thought and spoke about the problem in terms of static structures and discrete features like width-to-height ratios. These features do not easily offer a solution, and they led the participants to the wrong answer. Evidently, there are times when the verbal exchanges in a group lead to productive outcomes, and there are times when they do not.

By considering the type of understanding that people generate with the aid of language, it should be possible to prescribe particularly appropriate times to share linguistic representations during knowledge growth. For example, John Bransford and I have examined whether there is a "time for telling" (Schwartz & Bransford, in press). Is there a way to prepare people to be told something; are there times when the structured knowledge of verbal communication is particularly beneficial?

In one experiment, we tried to teach four target concepts from cognitive psychology (e.g. people tend to remember stereotypical events). Students were separated into three instructional treatments. In the "double-telling" treatment, students wrote a three-page summary of a brief book chapter that described the target concepts and the experiments that exemplified them. Five days later, they heard a lecture on the concepts and experiments. In the "double-discovery" treatment, students analyzed and looked for patterns in simplified data sets from experiments that exemplified the four concepts. Table 1 provides one sample of what they had to analyze. Five days later, they completed the analysis task again. In the final condition, "discovery and telling," the students analyzed the raw data and then, 5 days later, heard the same lecture as the "double-telling" students. A week after completing the instruction, all the students tried to predict the outcomes from an hypothetical study. All the target concepts were applicable to the hypothetical study. The relevant experimental question is which treatment made the most correct predictions?

The results were definitive. Students in the "discovery and telling" condition made over twice as many predictions as the students in either the "double-discovery" or "double-telling" conditions. One interpretation of this result is that the discovery activity helped students to discern specific features that differentiated classes of psychological phenomena, much as a botanist can distinguish sub-species of a given flower. For example, by analyzing Table 1, they differentiated events that have high and low frequencies in people's recounting of a visit to the doctor. After noticing this and other distinctions, the subsequent lecture provided the coherent and abstract structure that enabled these students to construct an understanding of why these differences are significant. For example, the lecture explained how schema theory predicts these types of stereotypy effects in memory.

Table 1. An example of the tasks that students completed to discover patterns in people's memory (Adapted from Schwartz & Bransford, in press).

*Researchers asked five people to write down the events that occur when they visit the doctor. The results are shown below. Analyze the data to discover the important patterns. Make a graph that shows those patterns.*

**Person 1:** Enter office. Check in with receptionist. Sit down. Wait. Name called. Enter exam room. Sit on table. Doctor examines. Doctor asks questions. Make another appointment. Leave office.

**Person 2:** Check in with receptionist. Read magazine. Look at other people. Name called. Sit on table. Nurse tests. Doctor examines. Leave office.

**Person 3:** Check in with receptionist. Sit down. Read magazine. Talk to nurse. Nurse tests. Talk to doctor about problem. Leave office.

**Person 4:** Enter office. Sit down. Read magazine. Enter exam room. Undress. Sit on table. Nurse tests. Doctor examines. Get dressed. Leave office.

**Person 5:** Enter office. Check in with receptionist. Sit down. Read magazine. Name called. Follow nurse. Enter exam room. Nurse tests. Doctor enters. Doctor examines.

The results of the preceding study show that individuals needed both forms of knowledge. Without the "discovery," the "telling" simply provided a set of facts to be memorized. And, without the "telling," the differentiations were simply observations. This result provides an important lesson for those who believe that direct teaching (e.g. a lecture) is contrary to constructivist ideals. As argued above, people construct their knowledge regardless of whether the input comes from the physical or linguistic world. The current study points out that there is a place for texts and lectures in a classroom; namely, when students have sufficiently differentiated domain knowledge to use the expository materials in a *constructive* manner.

In combination, the three studies indicate that it is useful to entertain which types of understanding people are most likely to construct under different learning conditions and how those types of understanding may complement one another. In particular, it appears that linguistic communications, whether within a collaborative group or a classroom lecture, lead to abstract and structured understandings. How well this hypothesis will hold up under further empirical scrutiny is an open question. Nonetheless, it seems important for researchers to consider seriously the type of knowledge that is most likely for people to construct in group interaction.

## 6 Computer Modeling Revisited

Ploetzner et al. state the following:

> "Even if a person models another person sufficiently to continue the dialogue, this might still be merely at a shallow level of understanding, not enough to actually learn." (p. 120, this volume)

The question then is how do we get people to move from shallow interactions to the deep interactions that can yield the special learning benefits of working in a group? As pointed out at the beginning of this chapter, little empirical research has yielded much headway on this problem. Thus, I have been trying to reconsider what is unique about the psychology of collaborative learning. To accomplish this task, I have worked before the backdrop of computer models of collaborative agents. Now, I would like to bring this backdrop into the foreground.

In the preceding sections, I have been laying out an agenda for computer models of collaborative learning by constructing a picture of what is essential to collaboration. I have argued that the computers must have the *agency* to choose to collaborate or not. Once they have that agency, they need *motivations* that determine whether they interact or not. Finally, they need to generate *learning* products that are likely to occur once individuals choose to cooperate. In the next three paragraphs, I briefly suggest some of the ways these ideas might be examined computationally. I am well beyond my depth here, so I may suggest things that have already been examined. I apologize in advance.

The way I have framed the issue of agency is by pitting choice versus social compliance. I think it would be very interesting to conduct simulations where one changes the balance of social rules and the individual agents' goals. One might program the social rules as highly rigid, as in the case of physical laws. Or, one might make them fairly soft, perhaps allowing agents to rebel against "society." At the same time, one might manipulate an agent's desire to maintain personal goals and to trust other agents. I wonder how changes in these various parameters would affect the interactions between the desire to comply, the desire to collaborate, and the desire to reach personal goals in a collaborative task. Artificial intelligence work in game playing seems particularly useful in this regard, because it has worked out methods for representing short-term and long-term personal goals, the goals of others, and the rules of the game. Perhaps it would be possible to construct a game-playing program where there are multiple agents playing multiple games, each trying to attain some form of points. The agents get to decide whether to "break the rules" in any given game, and they get to decide which agents they choose to play against and which agents they team up with. This would be a nice context for exploring the interplay among people's representations of other people's collaborativeness and their behavior with respect to social rules.

I framed the issue of motivation in terms of appropriation versus production. Moving this to the computational arena, one might create interesting simulations by developing "I will continue to collaborate" (ICC) heuristics. The ICC heuristics might take into account the degree to which the agent is appropriating another agent's knowledge, the degree to which the agent is producing knowledge, and the degree to which the agent perceives its productions as having been appropriated by another agent. By changing weights within the heuristics, it would be interesting to see if we could model outcomes from human interactions. Another possibility is that one might model what happens when one agent recognizes its ideas in another agent (i.e. when its ideas have been appropriated). For example, if agent A recognizes its "knowledge" in agent B, it may improve communication because agent A can rely on that knowledge as a common ground. Moreover, with some additional ICC heuristics, agent A might be more likely to communicate with agent B because it shows evidence of trying to make a shared meaning.

Finally, I framed the issue of learning in terms of assimilation versus construction. This is a more difficult problem for computer models, because they often rely on assimilations in which the propositions of one agent are directly inserted into the knowledge base of another agent. This problem could be avoided, however, by using inductive algorithms that treat propositional input as data from which to construct an understanding. More interesting in the current context, however, is the possibility of giving computer models different options for how they resolve breakdowns in communication. For example, will they try to find a physical referent so they can point to it, or will they generate a more abstract representation so they can communicate the structure of the idea they are having trouble communicating? More generally, I think Hoppe and Ploetzner (chapter 8, this volume) are on the right track when they look at multiple representations in the context of a qualitative and a quantitative agent that each try to communicate their knowledge. I wonder what would happen if they added a few more agents into the mix and gave them some ICC heuristics for choosing collaborators? Would quantitative agents end up associating with quantitative agents, and qualitative agents with qualitative agents?

One reason these considerations about collaboration are important, even as half-formed as they are, is that they raise the issue about the theoretical level at which we plan to operate. One approach would be to create computer models (or theories) that operate at the level of collaborative rules or roles. For example, one might manipulate the "collaborative rule sets" that different robots use and see what happens. These rules might take the form of Gricean maxims, "only communicate as much information as is necessary" or, they might take a more directly functional form, "exchange information if it will optimize the gathering of food." This would make for an interesting set of simulations, but they would be simulations about social conventions not about individuals in collaboration. Gricean maxims are only interesting because they can be broken. Moreover, as pointed out at the outset, Weiss and Dillenbourg (chapter 4, this volume) suggest that it may be difficult to prespecify all the needed rules because collaborative situations are often too complex; the agents need to be able to self-improve in their collaborative behaviors. This presumably requires a lower-order layer that can generate cooperative behaviors. I have been trying to propose some psychological elements of this lower layer. Ideally, what we want is a theory that explains how collaboration and collaborative learning emerge, not simply how they look once they appear. I believe analysis at this lower layer may best illuminate what is unique about people's effort to achieve shared meaning and deep understanding in small groups.

## Acknowledgements

I would like to thank John Bransford for the many insights he contributed directly and indirectly. I would also like to thank Pierre Dillenbourg for the opportunity to participate and Trefor Davies for his clever artwork in Figure 1.

# References

## Chapter 1

Amigues, R. (1987). Conceptual change and peer interaction. *Paper presented at the Second European Conference for Research on Learning and Instruction* (Tübingen, West Germany, September 19–22).

Baker, M.J. (1997). Structuring and scaffolding reflective interactions in a computer-supported collaborative learning environment. Invited Symposium on Tools and Interactions in Distributed Cognitive Systems. *Proceedings of the European Conference for Research on Learning and Instruction* (Athens, August).

Baker, M.J., & Lund, K. (1996). Flexibly structuring the interaction in a CSCL environment. In P. Brna, A. Paiva, & J. Self (Eds.), *Proceedings of the European Conference on Artificial Intelligence in Education* (Lisbon, Portugal, September 20–October 2) (pp. 401–407).

Blaye, A. (1988). Confrontation socio-cognitive et résolution de problèmes. Doctoral dissertation, Centre de Recherche en Psychologie Cognitive, Université de Provence, Aix-en-Provence, France.

Chi M.T.H., Bassok, M., Lewis, M.W., Reimann, P., & Glaser, R. (1989). Self-explanations: How students study and use examples in learning to solve problems. *Cognitive Science, 13*, 145–182.

Clark, H.H., & Brennan S.E. (1991). Grounding in communication. In L. Resnick, J. Levine, & S. Teasley (Eds.), *Perspectives on Socially Shared Cognition* (pp. 127–149). Hyattsville, MD: American Psychological Association.

Dillenbourg, P., & Baker, M (1996). Negotiation spaces in human–computer collaborative learning. In *Proceedings of the International Conference on Cooperative Systems (COOP'96)* (Juan-les-Pins, France, June 12–14).

Dillenbourg, P., & Traum, D. (1996). Grounding in multi-modal task-oriented collaboration. In P. Brna, A. Paiva, & J. Self (Eds.), *Proceedings of the European Conference on Artificial Intelligence in Education* (Lisbon, Portugal, September 20–October 2) (pp. 415–425).

Dillenbourg, P., Baker, M., Blaye, A., & O'Malley, C. (1995). The evolution of research on collaborative learning. In E. Spada, & P. Reiman (Eds.), *Learning in Humans and Machines: Towards an Interdisciplinary Learning Science* (pp. 189–211). Oxford: Elsevier.

Dillenbourg, P., Jermann, P., Schneider D., Traum, D., & Buiu, C. (1997). The design of MOO agents: Implications from an empirical CSCW study. In *Proceedings of the 8th World Conference on Artificial Intelligence in Education* (Kobe, Japan). (pp. 15–22).

Doise, W., & Mugny, G. (1984). *The Social Development of the Intellect*. Oxford: Pergamon.

Doise, M., Mugny, G., & Perret-Clermont, A.-N. (1975) Social interactions and the development of cognitive operations. *European Journal of Social Psychology, 5*, 367–383.

Douglas, S.A. (1991). Tutoring as interaction: Detecting and repairing tutoring failures. In P. Goodyear (Ed.), *Teaching Knowledge and Intelligent Tutoring* (pp. 123–148). Norwood, NJ: Ablex.

Durfee, E.H., Lesser, V.R., & Corkill, D.D. (1989). Cooperative distributed problem solving. In A. Barr, P.R. Cohen, & E.A. Feigenbaum (Eds.), *The Handbook of Artificial Intelligence* (Vol. IV, pp. 83–127). Reading, MA: Addison-Wesley.

Fox, B. (1987). Interactional reconstruction in real-time language processing. *Cognitive Science, 11*(3), 365–387.

Gasser, L., & Huhns, M.N. (1989). *Distributed Artificial Intelligence, Vol. II.* London: Pitman.

Gilly, M. (1989). The psychosocial mechanisms of cognitive constructions, experimental research and teaching perspectives. *International Journal of Educational Research, 13*(6), 607–621.

Hutchins, E. (1995). How a cockpit remembers its speeds. *Cognitive Science, 19*, 265–288.

Jermann, P., & Schneider, D.K. (1997). Semi-structured interface in collaborative problem-solving. *Swiss Workshop on Collaborative and Distributed Systems* (Lausanne, Switzerland, May 1).

Krammer, H.P.M., & Van Merriënboer, J.J.G. (1989). Causal relationships between group heterogeneity, problem-solving activities, verbal interaction and learning outcomes in an introductory programmeming course. Internal Report, Department of Instructional Psychology, University of Twente, Enschede, The Netherlands.

Ligorio, B. (1997). Social influence in a text-based virtual reality. Unpublished master's thesis in social psychology, School of Education and Psychology, University of Geneva, Switzerland.

Minsky, M. (1987). *The Society of Mind.* London: William Heinemann.

Mitchell, T.M., Keller, R.M., & Kedar-Cabelli, S.T. (1986). Explanation-based generalization: A unifying view. *Machine Learning, 1*(1), 47–80.

Miyake, N. (1986). Constructive interaction and the iterative process of understanding. *Cognitive Science, 10*, 151–177.

Müller, J.P., & Rodriguez, M. (1995). Representation and planning for behavior-based robots. In H. Bunke et al. (Eds.), *Modelling and Planning for Sensor Based Intelligent Robot Systems.* River Edge, NJ: World Scientific.

O'Conaill, B., Whittaker, S., & Wilbur S. (1993). Conversations over video conferences: An evaluation of the spoken aspects of video-mediated communication. *Human–Computer Interaction, 8*(4), 389–428.

Palincsar, A.S., & Brown, A.L. (1984). Reciprocal teaching of comprehension-fostering and comprehension-monitoring activities. *Cognition and Instruction, 1*(2), 117–175.

Pea, R.D. (1993). Learning scientific concepts through material and social activities: Conversational analysis meets conceptual change. *Educational Psychologist, 28*(3), 265–277.

Perkins, D.N. (1993). Person-plus: A distributed view of thinking and learning. In G. Salomon (Ed.), *Distributed Cognitions. Psychological and Educational Considerations* (pp. 88–110). Cambridge, MA: Cambridge University Press.

Perret-Clermont, A.-N., Perret J.-F., & Bell, N. (1991). The social construction of meaning and cognitive activity in elementary school children. In L. Resnick, J. Levine, & S. Teasley (Eds.), *Perspectives on Socially Shared Cognition* (pp. 41–62). Hyattsville, MD: American Psychological Association.

Rogoff, B. (1990). *Apprenticeship in Thinking.* New York: Oxford University Press.

Roschelle, J. (1992). Learning by collaborating: Convergent conceptual change. *Journal of the Learning Sciences, 2*, 235–276.

Roschelle, J., & Teasley, S.D. (1995). The construction of shared knowledge in collaborative problem solving. In C.E. O'Malley (Ed.), *Computer-Supported Collaborative Learning* (pp. 69–197). Berlin: Springer-Verlag.

Salomon, G. (1993). No distribution without individual's cognition. A dynamic interactional view. In G. Saloman (Ed.), *Distributed Cognitions. Psychological and Educational Considerations* (pp. 111–138). Cambridge, MA: Cambridge University Press.
Schnotz, W. (1997). Individual and co-operative acquisition of knowledge with static and animated pictures in computer-based learning environments. *Paper presented at the European Conference on Learning and Instructions* (Athens, August).
Schwartz, D.L. (1995). The emergence of abstract dyad representations in dyad problem solving. *Journal of the Learning Sciences*, *4*(3), 321–354.
VanLehn, K., Jones, R.M., & Chi, M.T.H. (1992). A model of the self-explanation effect. *Journal of The Learning Sciences*, *1*, 1–59.
Webb, N.M (1989). Peer interaction and learning in small groups. *International Journal of Educational Research*, *13*(1), 21–39.
Webb, N.M. (1991). Task related verbal interaction and mathematical learning in small groups. *Research in Mathematics Education*, *22*(5), 366–389.
Weiss, G. (Ed.). (1997). *Distributed Artificial Intelligence Meets Machine Learning* (Lecture Notes in Artificial Intelligence). Berlin: Springer-Verlag.
Wertsch, J.V. (1985). Adult–child interaction as a source of self-regulation in children. In S.R. Yussen (Ed.), *The Growth of Reflection in Children* (pp. 69–97). Madison, WI: Academic Press.

## Chapter 2

Barfurth, M.A. (1995). Understanding the collaborative learning process in a technology rich environment: The case of children's disagreements. In J.L. Schnase, & E.L. Cunnius (Eds.), *Proceedings of the First International Conference on Computer Support for Collaborative Learning* (pp. 8–13). Bloomington, IN: Lawrence Erlbaum Associates.
Blaye, A. (1988). Confrontation sociocognitive et resolution de probleme. Unpublished doctoral dissertation, University of Provence, Aix-en-Provence, France.
Brown, J., Collins, A., & Duguid, P. (1989). Debating the situation: A rejoinder to Palincsar and Wineburg. *Educational Researcher*, *18*(4), 10–12.
Collins, A., Brown, J.S., & Newman, S.E. (1989). Cognitive apprenticeship: Teaching the crafts of reading, writing, and mathematics. In L.B. Resnick (Ed.), *Knowing, Learning and Instruction: Essays in Honor of Robert Glaser* (pp. 453–494). Hillsdale, NJ: Lawrence Erlbaum Associates.
Crook, C. (1996). *Computers and the Collaborative Experience of Learning*. London: Routledge.
Dillenbourg, P., Baker, M., Blaye, A., & O'Malley, C. (1995). The evolution of research on collaborative learning. In H. Spada, & P. Reimann (Eds.), *Learning in Humans and Machines* (pp. 189–211). Oxford: Elsevier.
Doise, W., & Mugny, G. (1984). *The Social Development of the Intellect*. Oxford: Pergamon.
Forman, E.A., & Cazden, C.B. (1985). Exploring Vygotskian perspectives in education: The cognitive value of peer interaction. In J. Wertsch (Ed.), *Culture, Communication, and Cognition: Vygotskian Perspectives* (pp. 323–347). Cambridge: Cambridge University Press.
Forman, E., & Larreamendy-Joerns, J. (1995). Learning in the context of peer collaboration: A pluralistic perspective on goals and expertise. *Cognition and Instruction*, *13*, 549–564.
Howe, C., Tolmie, A., & Anderson, A. (1991). Information technology and groupwork in physics. *Journal of Computer Assisted Learning*, *7*, 133–143.

Howe, C., Tolmie, A., Anderson, A., & MacKenzie, M. (1992). Conceptual knowledge in physics: The rôle of group interaction in computer supported teaching. *Learning and Instruction, 2*, 161–183.

Järvelä, S. (1995). The cognitive apprenticeship model in a technologically rich learning environment: Interpreting the learning interaction. *Learning and Instruction, 5*(3), 237–259.

Joiner, R., Messer, D., Light, P., & Littleton, K. (1995). Peer interaction and peer presence in computer-based problem-solving: A research note. *Cognition and Instruction, 13*(4), 583–584.

Keogh, T., & Barnes, P. (1994). Computers, verses, paper: Girls versus boys. *Paper presented at the International Conference of Group and Interactive Learning* (University of Strathclyde, September).

Koschmann, T.D. (1994). Toward a theory of computer support for collaborative learning. *Journal of the Learning Sciences, 3*(3), 219–225.

Lave, J., & Wenger, E. (1991). *Situated Learning: Legitimate Peripheral Participation.* Cambridge: Cambridge University Press.

Light, P., & Glachan, M. (1985). Facilitation of problem solving through peer interaction. *Educational Psychology, 5*, 217–225.

Light, P., & Woodhead, M. (1991). *Child Development in Social Context: Study Guide.* Milton Keynes: The Open University.

Light, P., Littleton, K., Messer, D., & Joiner, R. (1994). Social and communicative processes in computer-based problem solving. *European Journal of Psychology of Education, 9*(1), 93–109.

Mercer, N. (1995). *The Guided Construction of Knowledge: Talk Amongst Teachers and Learners.* Clevedon: Multilingual Matters.

Mercer, N., & Fisher, E. (1992). How do teachers help children to learn? An analysis of teachers' interventions in computer-based activities. *Learning and Instruction, 2*, 339–355.

Mercer, N., & Wegerif, R. (in press). Is exploratory talk productive talk? In K. Littleton, & P. Light (Eds.), *Learning with Computers: Analysing Productive Interaction.* London: Routledge.

Monteil, J.-M. (1991). Social regulation and individual cognitive function: The effects of individuation on cognitive performance. *European Journal of Social Psychology, 21*, 225–237.

Monteil, J.-M. (1992). Toward a social psychology of cognitive functioning: Theoretical outline and empirical illustrations. In M. von Granach, W. Doise, & G. Mugny (Eds.), *Social Representations and the Social Bases of Knowledge* (pp. 94–105). Berne: Hubert.

Nystrand, M. (1986). *The Structure of Written Communication: Studies of Reciprocity Between Writers and Readers.* London: Academic Press.

Pea, R.D. (1994). Seeing what we build together: Distributed multimedia learning environments for transformative communications. *Journal of the Learning Sciences, 3*(3), 285–299.

Perret-Clermont, A.-N. (1980). *Social Interaction and Cognitive Development in Children.* London: Academic Press.

Piaget, J. (1932). *The Moral Development of the Child.* London: Routledge.

Robinson-Stavely, K., & Cooper, J. (1990). Mere presence, gender and reactions to computers. *Journal of Experimental Social Psychology, 26*, 168–183.

Rogoff, B. (1990). *Apprenticeship in Thinking: Cognitive Development in Social Context.* Oxford: Oxford University Press.

Säljö, R. (1995). Mental and physical artifacts in cognitive practices. In P. Reimann, & H. Spada (Eds.), *Learning in Humans and Machines. Towards an Interdisciplinary Learning Science* (pp. 83–96). Oxford: Pergamon.

Salomon, G. (1994). Differences in patterns: Studying computer enhanced learning environments. In S. Vosniadu, E. De Corte, & H. Mandl (Eds.), *Technology-based Learning Environments: Psychological and Educational Foundations*. (NATO ASI Series F: Computer and System Sciences, Vol. 137) (pp. 70–85). Berlin: Springer.

Schoenfeld, A.H. (1992). On paradigms and methods: What do you do when the ones you know don't do what you want them to? Issues in the analysis of data in the form of videotapes. *Journal of the Learning Sciences*, *2*(2), 179–214.

Sternberg, R.J., & Wagner, R.K. (Eds.). (1994). *Mind in Context: Interactionist Perspectives on Human Intelligence*. Cambridge: Press Syndicate of the University of Cambridge.

Suchman, L. (1987). *Plans and Situated Actions*. Cambridge: Cambridge University Press.

Teasley, S., & Roschelle, J. (1993). Constructing a joint problem space: The computer as a tool for sharing knowledge. In S.P. Lajoie, & S.J. Derry (Eds.), *Computers as Cognitive Tools* (pp. 229–257). Hillsdale, NJ: Lawrence Erlbaum Associates.

Tudge, J. (1989). When collaboration leads to regression: Some negative consequences of socio-cognitive conflict. *European Journal of Social Psychology*, *19*, 123–138.

Tudge, J., & Rogoff, B. (1989). Peer influences on cognitive development: Piagetian and Vygotskian perspectives. In M.H. Bornstein, & J.S. Bruner (Eds.), *Interaction in Human Development* (pp. 17–40). Hillsdale, NJ: Lawrence Erlbaum Associates.

Vosniadou, S. (1996). Towards a revised cognitive psychology for new advances in learning and instruction. *Learning and Instruction*, *6*(2), 95–109.

Vygotsky, L.S. (1978). *Mind in Society*. Cambridge, MA: Harvard University Press.

Wertsch, J.V. (1991). A socio-cultural approach to socially shared cognition. In L.B. Resnick, J.M. Levine, & S.D. Teasley (Eds.), *Perspectives on Socially Shared Cognition* (pp. 1–20). Washington, DC: American Psychological Association.

## Chapter 3

Agre, P. (1995). Computational research on interaction and agency. In P. Agre, & S. Rosenschein (Eds.), (Computational Theories of Interaction and Agency Special Issue). *Artificial Intelligence*, *72*, 1–52.

Allen, J.F., Schubert, L.K., Ferguson, G., Heeman, P., Hwang, C. H., Kato, T., Light, M., Martin, N., Miller, B., Poesio, M., & Traum, D.R. (1994). The TRAINS project: A case study in building a conversational planning agent. *Journal of Experimental and Theoretical Artificial Intelligence*, *7*, 7–48.

Allwood, J. (1976). *Linguistic Communication in Action and Co-operation: A Study in Pragmatics* (Gothenburg Monographs in Linguistics 2). Sweden: University of Gothenburg, Department of Linguistics.

Allwood, J., Nivre, J., & Ahlsén, E. (1991). *On the Semantics and Pragmatics of Linguistic Feedback* (Gothenburg Papers in Theoretical Linguistics No. 64). Sweden: University of Gothenburg, Department of Linguistics.

Baker, M.J. (1994). A model for negotiation in teaching-learning dialogues. *Journal of Artificial Intelligence in Education*, *5*(2), 199–254.

Baker, M.J. (1995). Negotiation in collaborative problem-solving dialogues. In R.-J. Beun, M.J. Baker, & M. Reiner (Eds.), *Dialogue and Instruction* (pp. 39–55). Berlin: Springer-Verlag.

Baker, M.J. (1996). Argumentation et co-construction des connaissances. *Interaction et Cognitions*, *1*(2/3), 157–191.

Baker, M.J., & Lund, K. (1997). Promoting reflective interactions in a CSCL environment. *Journal of Computer Assisted Learning*, *13*, 175–193.

Bunt, H.C. (1995). Dialogue control functions and interaction design. In R.J. Beun, M.J. Baker, & M. Reiner (Eds.), *Dialogue and Instruction* (pp. 197–214). Berlin: Springer-Verlag.

Clark, H.H. (1994). Managing problems in speaking. *Speech Communication*, *15*, 243–250.

Clark, H.H., & Brennan, S. (1991). Grounding in communication. In L.B. Resnick, J.M. Levine, & S.D. Teasley (Eds.), *Perspectives on Socially Shared Cognition* (pp. 127–149). Washington, DC: American Psychological Association.

Clark, H.H., & Marshall, C.R. (1981). Definite reference and mutual knowledge. In A.K. Joshi, B.L. Webber, & I.A. Sag (Eds.), *Elements of Discourse Understanding* (pp. 10–63). Cambridge: Cambridge University Press.

Clark, H.H., & Schaefer, E.F. (1987). Collaborating on contributions to conversation. *Language and Cognitive Processes*, *2*, 1–23.

Clark, H.H., & Schaefer, E.F. (1989). Contributing to discourse. *Cognitive Science*, *13*, 259–294.

Clark, H.H., & Wilkes-Gibbs, D. (1986). Referring as a collaborative process. *Cognition*, *22*, 1–39.

Cole, M., & Engeström, Y. (1993). A cultural–historical approach to distributed cognition. In G. Salomon (Ed.), *Distributed Cognitions: Psychological and Educational Considerations* (pp. 1–46). Cambridge, MA: Cambridge University Press.

Danieli, M., & Gerbino, E. (1995). Metrics for evaluating dialogue strategies in a spoken language system. In *Proceedings of the 1995 AAAI Spring Symposium on Empirical Methods in Discourse Interpretation and Generation* (pp. 34–39).

Devi, R., Tiberghien, A., Baker, M., & Brna, P. (1996). Modelling students' construction of energy models in physics. *Instructional Science*, *24*, 259–293.

Dillenbourg, P., Jermann, D., Schneider, D., Traum, D., & Buiu, C. (1997). The design of MOO agents: Implications from a study on multi-modal collaborative problem solving. In *Proceedings of the 8th World Conference on Artificial Intelligence in Education (AI-ED 97)* (Kobé, Japan) (pp. 15–22). Amsterdam: IOS Press.

Forman, E.A. (1992). Discourse, intersubjectivity, and the development of peer collaboration: A Vygotskian approach. In L.T. Winegar, & J. Valsiner (Eds.), *Children's Development within Social Context: Vol 2. Metatheoretical, Theoretical and Methodological Issues* (pp. 143–159). Hillsdale, NJ: Lawrence Erlbaum Associates.

Johnson-Laird, P.N. (1988). *The Computer and the Mind: An Introduction to Cognitive Science*. London: Fontana.

Lave, J. (1991). Situating learning in communities of practice. In L.B. Resnick, J.M. Levine, & S.D. Teasley (Eds.), *Perspectives on Socially Shared Cognition* (pp. 63–84). Washington DC: American Psychological Association.

Leontjev, A.N. (1981). The problem of activity in psychology. In J.V. Werstch (Ed.), *The Concept of Activity in Soviet Psychology* (pp. 37–71). Armonk, NY: Sharp.

Levinson, S.C. (1983). *Pragmatics*. Cambridge: Cambridge University Press.

Luria, A.R. (1979). *The Making of Mind*. Cambridge, Mass.: Harvard University Press.

Mammen, J. (1993). The elements of psychology. In N. Engelsted, M. Hedegaard, B. Karpatschof, & A. Mortensen (Eds.), *The Societal Subject* (pp. 29–44). Aarhus: Aarhus University Press.

Miyake, N. (1986). Constructive interaction and the iterative process of understanding. *Cognitive Science*, *10*(2), 151–177.

Moeschler, J. (1985). *Argumentation et conversation*. Paris: Crédif-Hatier.

Perret-Clermont, A.-N., Perret, J.-F., & Bell, N. (1991). The social construction of meaning and cognitive activity in elementary school children. In L.B. Resnick, J.M. Levine, & S.D.

Teasley (Eds.), *Perspectives on Socially Shared Cognition* (pp. 41–62). Washington DC: American Psychological Association.

Rommetveit, R. (1979). On the architecture of intersubjectivity. In R. Rommetveit, & R. M. Blaker (Eds.), *Studies of Language, Thought, and Verbal Communication* (pp. 93–108). London: Academic Press.

Rommetveit, R. (1985). Language acquisition as increasing linguistic structuring of experience and symbolic behaviour control. In J. Wertsch (Ed.). *Culture Communication and Cognition: Vygotskian Perspectives* (pp. 183–204). Cambridge: Cambridge University Press.

Roschelle, J., & Teasley, S.D. (1995). The construction of shared knowledge in collaborative problem solving. In C. O'Malley (Ed.), *Computer Supported Collaborative Learning* (pp. 69–100). Berlin: Springer-Verlag.

Roulet, E. (1992). On the structure of conversation as negotiation. In H. Parret, & J. Verschueren (Eds.), *(On) Searle on Conversation* (pp. 91–100). Amsterdam: John Benjamins.

Schwartz D.L. (1995). The emergence of abstract representations in dyadic problem solving. *Journal of the Learning Sciences*, *4*, 321–354.

Smith, R.W., & Hipp, D.R. (1994). *Spoken Natural Language Dialog Systems: A Practical Approach*. Oxford: Oxford University Press.

Tiberghien, A. (1994). Modeling as a basis for analyzing teaching–learning situations. *Learning and Instruction*, *4*, 71–87.

Tiberghien, A. (1996). Construction of prototypical situations in teaching the concept of energy. In G. Welford, J. Osborne, & P. Scott (Eds.), *Research in Science Education in Europe* (pp. 100–114). London: Falmer Press.

Tolman, C.W. (1988). The basic vocabulary of activity theory. *Multidisciplinary Newsletter for Activity Theory*, *1*, 14–20.

Traum, D.R. (1994). A computational theory of grounding in natural language conversation. Ph.D. thesis, Department of Computer Science, University of Rochester. Also available as TR 545, Department of Computer Science, University of Rochester.

Vygotsky, L. (1978). *Mind in Society: The Development of Higher Psychological Processes*. [ed. M. Cole, V. John-Steiner, S. Scribner, & E. Souberman], Cambridge: Cambridge University Press.

Vygotsky, L. (1986). *Thought and Language*. [ed. A. Kozulin], Cambridge, MA: MIT Press.

Wells, G. (1996). Using the tool-kit of discourse in the activity of learning and teaching. *Mind, Culture, and Activity*, *3*(2), 74–101.

Werstch, J.V. (1985). *Vygotsky and the Social Formation of Mind*. Cambridge, MA: Harvard University Press.

Wertsch, J.W. (1994). The primacy of mediated action in sociocultural studies. *Mind, Culture, and Activity*, *1*(4), 202–208.

Winograd, T., & Flores, F. (1986). *Understanding Computers and Cognition*. Reading, MA: Addison-Wesley.

## Chapter 4

Baker, M. (1992). The collaborative construction of explanations. *Paper presented at Deuxièmes Journées Explication du PRC-GDR-IA du CNRS* (Sophia-Antipolis, June 17–19).

Bazzan, A. (1997). Evolution of coordination as a metaphor for learning in multi-agent systems. In G. Weiss (Ed.), *Distributed Artificial Intelligence Meets Machine Learning* (Lecture Notes in Artificial Intelligence, Vol. 1221) (pp. 117–136). Berlin: Springer–Verlag.

Benda, M., Jaganathan, V., & Dodhiawala, R. (1986). *On Optimal Cooperation of Knowledge Sources*. Technical Report. Seattle, WA: Boing Advanced Technical Center, Boing Computer Services.

Blaye, A. (1988). Confrontation socio-cognitive et resolution de problemes. Doctoral dissertation, Centre de Recherche en Psychologie Cognitive, Université de Provence, Aix-en-Provence, France.

Bond, A.H., & Gasser, L. (Eds.) (1988). *Readings in Distributed Artificial Intelligence*. San Mateo, CA: Morgan Kaufmann.

Bui, H.H., Kieronska, D., & Venkatesh, S. (1996). Negotiating agents that learn about others' preferences. In S. Sen (Ed.), *Adapation, Coevolution and Learning in Multiagent Systems* (Papers from the 1996 AAAI Symposium) Technical Report SS-96-01. (pp. 16–21). Palo Alto, CA: AAAI Press.

Bull, L., & Fogarty, T. (1996). Evolutionary computing in cooperative multi-agent environments. In S. Sen (Ed.), *Adapation, Coevolution and Learning in Multiagent Systems* (Papers from the 1996 AAAI Symposium) Technical Report SS-96-01. (pp. 22–27). Palo Alto, CA: AAAI Press.

Carmel, D., & Markovitch, S. (1996). Opponent modeling in multi-agent systems. In G. Weiss, & S. Sen (Eds.), *Adaptation and Learning in Multi-agent Systems* (Lecture Notes in Artificial Intelligence, Vol. 1042) (pp. 40–52). Berlin: Springer-Verlag.

Chan, P.K., & Stolfo, S.J. (1993). Toward parallel and distributed learning by meta-learning. *Working Notes of the AAAI Workshop on Knowledge Discovery and Databases* (pp. 227–240).

Chi, M.T.H., Bassok, M., Lewis, M.W., Reimann, P., & Glaser, R. (1989). Self-explanations: How students study and use examples in learning to solve problems. *Cognitive Science*, *13*, 145–182.

Dillenbourg P., & Schneider D. (1995). Mediating the mechanisms which make collaborative learning sometimes effective. *International Journal of Educational Telecommunications*, *1*(2/3), 131–146.

Dillenbourg, P., & Self, J.A. (1992). A computational approach to socially distributed cognition. *European Journal of Psychology of Education*, *3*(4), 353–372.

Dillenbourg, P., Jermann, P. , Buiu, C., Traum, D., & Schneider D. (1997). The design of MOO agents: Implications from an empirical CSCW study. In *Proceedings of the 8th World Conference on Artificial Intelligence in Education* (Kobe, Japan).

Doise, W., & Mugny, G. (1984). *The Social Development of the Intellect*. Oxford: Pergamon.

Fischer, K., Kuhn, N., Müller, H.J., Müller, J.P., & Pischel, M. (1993). Sophisticated and distributed: The transportation domain. In *Proceedings of the Fifth European Workshop on Modelling Autonomous Agents in a Multi-Agent World (MAAMAW-93)*.

Gasser, L., & Huhns, M.N. (Eds.), (1989). *Distributed Artificial Intelligence, Vol. 2*. London: Pitman.

Grefenstette, J., & Daley, R. (1996). Methods for competitive and cooperative co-evolution. In S. Sen (Ed.), *Adapation, Coevolution and Learning in Multiagent Systems* (Papers from the 1996 AAAI Symposium) Technical Report SS-96-01. (pp. 45–50) Palo Alto, CA: AAAI Press.

Haynes, T., & Sen, S. (1996). Evolving behavioral strategies in predators and prey. In G. Weiss, & S. Sen (Eds.), *Adaptation and Learning in Multi-agent Systems* (Lecture Notes in Artificial Intelligence, Vol. 1042) (pp. 113–126). Berlin: Springer–Verlag.

Haynes, T., Lau, K., & Sen, S. (1996). Learning cases to compliment rules for conflict resolution in multi-agent systems. In S. Sen (Ed.), *Adapation, Coevolution and Learning in Multi-agent Systems* (Papers from the 1996 AAAI Symposium) Technical Report SS-96-01 (pp. 51–56). Palo Alto, CA: AAAI Press

Huhns, M.N. (Ed.). (1987). *Distributed Artificial Intelligence*. London: Pitman.

Imam, I.F. (Ed.). (1996). *Intelligent Adaptive Agents*. (Papers from the 1996 AAAI Workshop.) Technical Report WS-96–04. Palo Alto, CA: AAAI Press.

Mataric, M.J. (1996). Learning in multi-robot systems. In G. Weiss, & S. Sen (Eds.), *Adaptation and Learning in Multi-agent Systems* (Lecture Notes in Artificial Intelligence, Vol. 1042) (pp. 152–163). Berlin: Springer-Verlag.

Michalski, R., & Tecuci, G. (Eds.), (1995). *Machine Learning. A Multistrategy Approach*. San Mateo, CA: Morgan Kaufmann.

Mitchell, T.M., Keller, R.M., & Kedar-Cabelli S.T. (1986). Explanation-based generalization: A unifying view. *Machine Learning*, *1*(1), 47–80.

Miyake, N. (1986). Constructive interaction and the iterative process of understanding. *Cognitive Science*, *10*, 151–177.

Mor, Y., Goldman, C.V., & Rosenschein, J.S. (1996). Learn your opponent's strategy (in polynomial time)! In G. Weiss, & S. Sen (Eds.), *Adaptation and Learning in Multi-agent Systems* (Lecture Notes in Artificial Intelligence, Vol. 1042) (pp. 164–176). Berlin: Springer-Verlag.

Müller, J., & Pischel, M. (1994). An architecture for dynamically interacting agents. *Journal of Intelligent and Cooperative Information Systems*, *3*(1), 25–45.

Müller, J., Wooldridge, M., & Jennings, N. (Eds.) (1997). *Intelligent Agents III* (Lecture Notes in Artificial Intelligence, Vol. 1193). Berlin: Springer-Verlag.

Nagendra Prasad, M.V., Lesser, V.R., & Lander, S.E. (1995). *Learning Organizational Roles in a Heterogeneous Multi-agent System*. Technical Report 95–35. Amherst, MA: Computer Science Department, University of Massachusetts.

Nagendra Prasad, M.V., Lesser, V.R., & Lander, S. (1996). On reasoning and retrieval in distributed case bases. (Digital Libraries Special Issue.) *Journal of Visual Communication and Image Representation*, *7*(1), 74–87.

O'Hare, G.M.P., & Jennings, N.R. (Eds.), (1996). *Foundations of Distributed Artificial Intelligence*. New York: John Wiley, & Sons.

Ohko, T., Hiraki, K., & Anzai, Y. (1996). Learning to reduce communication cost on task negotiation among multiple autonomous mobile robots. In G. Weiss, & S. Sen (Eds.), *Adaptation and Learning in Multi-agent Systems* (Lecture Notes in Artificial Intelligence, vol. 1042) (pp. 177–191). Berlin: Springer-Verlag.

Parker, L. (1993). Adaptive action selection for cooperative agent teams. In *Proceedings of the Second International Conference on Simulation of Adaptive Behavior* (pp. 442–450).

Peceny, M., Weiss, G., & Brauer, W. (1996). *Verteiltes maschinelles Lernen in Fertigungsumgebungen*. Report FKI-218–96. Munich: Institut für Informatik, Technische Universität München.

Plaza, E., Arcos, J.L., & Martin, F. (1997). Cooperative case-based reasoning. In G. Weiss (Ed.), *Distributed Artificial Intelligence Meets Machine Learning* (Lecture Notes in Artificial Intelligence, Vol. 1221) (pp. 180–201). Berlin: Springer-Verlag.

*Proceedings of the First International Conference on Multi-Agent Systems (ICMAS-95)* (1995). Palo Alto, CA/Cambridge, MA: AAAI Press/MIT Press.

*Proceedings of the Second International Conference on Multi-Agent Systems (ICMAS-96)* (1996). Palo Alto, CA/Cambridge, MA: AAAI Press/MIT Press

# Chapter 5


Amigue, R. (1989). Peer interaction and conceptual change. In H. Mandle, E. De Corte, N. Bennett, & H.F. Freidrich (Eds.), *Learning and Instruction: European Research in an International Context* (Vol. 2, pp. 27–43). Oxford: Pergamon.

Amigue, R., & Agostinelli, S. (1992). Collaborative problem solving with a computer: How can an interactive learning environment be designed? *European Journal of Psychology of Education*, *7*(4), 325–337.

Balch, T., & Arkin, R. (1994). *Communication in Reactive Multiagent Robotic Systems*. Technical Report GIT-CC-95–47. Atlanta, GA: Georgia Institute of Technology.

Barbieri, M.S., & Light, P.H. (1992). Interaction, gender and performance on a computer-based problem solving task. *Learning and Instruction*, *2*, 199–213.

Bennett, N., & Dunne, E. (1991). The nature and quality of talk in co-operative classroom groups. *Learning and Instruction*, *1*, 103–118.

Blaye, A., Light, P., Joiner, R., & Sheldon, S. (1991). Collaboration as a facilitator of planning and problem solving on a computer based task. *British Journal of Developmental Psychology*, *9*, 471–483.

Brooks, R., & Flynn, A. (1989). *Fast, Cheap, and Out of Control*, AI Memo 1182. Massachusetts: AI Laboratory, MIT.

Chan, T.-W., & Baskin, A.B. (1990). Learning companion systems. In C. Frasson, & G. Gauthier (Eds.), *Intelligent tutoring systems: At the Crossroads of Artificial Intelligence and Education*. Norwood, NJ: Ablex.

Chan, T.-W., & Chou, C.-Y. (1995). Simulating a learning companion in reciprocal tutoring systems. In J.L. Schnase, & E.L. Cunnius (Eds.), *Proceedings of CSCL 95*. Mahwah, NJ: Lawrence Erlbaum Associates.

Clements, D.H., & Nastasi, B.K. (1988). Social and cognitive interactions in educational computer environments. *American Educational Research Journal*, *25*(1), 87–106.

Cognition and Technology Group at Vanderbilt (1992). The Jasper series as an example of anchored instruction: Theory program description and assesment data. *Educational Psychologist*, *27*(3), 291–315.

Crook, C. (1987). Computers in the classroom: defining a social context. In J. Rutkowska, & C. Crook (Eds.), *Computers, Cognition and Development*. Chichester: Wiley.

Crook, C. (1994). *Computers and the Collaborative Experience of Learning*. London: Routledge.

Dalton, D.W. (1990). The effects of cooperative learning strategies on achievement and attitudes during interactive video. *Journal of Computer Based Instruction*, *17*, 8–16.

Deneubourg, J.L., Gross, S., Franks, N., Sendova-Franks, A., Detrain, C., and Chretien, L. (1991). The dynamics of collective sorting: Robot-like ants and ant-like robots. In *Proceedings of the First International Conference on Simulation of Adaptive Behaviour*. Cambridge, MA: MIT Press.

Dillenbourg, P., & Self, J. (1992). A computational approach to socially distributed cognition. *European Journal of Psychology of Education*, *7*(4), 353–372.

Drogoul, A., & Ferber, J. (1992). From Tom Thumb to the dockers: Some experiments with foraging robots. In *Proceedings of the Second International Conference on Simulation of Adaptive Behaviour*. Cambridge, MA: MIT Press.

Foot, H.C., Howe, C.J., Anderson, A., Tolmie, A. K., & Warden, D.A. (1994). *Group and Interactive Learning*. Ashurst, Kent: Computational Mechanics Publications.

Healey, L., Pozzi, S., & Hoyles C. (1995). Making sense of groups, computers and mathematics. *Cognition and Instruction*, *13*(4), 505–524.

Paljung, E., & Yun, X. (1993). Experimental results of two robot arms manipulating large objects. In *Proceedings of the International Conference on Robotics and Automation (ICRA93)*, Atlanta, GA: IEEE Press.

Parker, L.E. (1992). Adaptive action selection for cooperative agent teams. In *Proceedings of the Second International Conference on Simulation of Adaptive Behaviour*. Cambridge, MA: MIT Press.

Pea, R.D. (1996) Seeing what we build together: Distributed multimedia learning environments for transformative communications. In T. Koschmann (Ed.), *CSCL: Theory and Practice of an Emerging Paradigm*. Mahwah, NJ: Lawrence Erlbaum Associates.

Pozzi, S., Healey, L., & Hoyles C. (1993). Learning and interaction in groups with computers: When do ability and gender matter? *Social Development*, *2*(3), 222–241.

Riel, M. (1996). Cross-classroom collaboration: Communication and education. In T. Koschmann (Ed.), *CSCL: Theory and Practice of an Emerging Paradigm*. Mahwah, NJ: Lawrence Erlbaum Associates.

Roschelle, J., & Teasley, S.D. (1995). The construction of shared knowledge in collaborative problem solving. In C. O'Malley (Ed.), *Computer Supported Collaborative Learning*. Berlin: Springer-Verlag.

Scardamalia, M., & Bereiter, C. (1996). Computer support for knowledge building. In T. Koschmann (Ed.), *CSCL: Theory and Practice of an Emerging Paradigm*. Mahwah, NJ: Lawrence Erlbaum Associates

Siann, G., & Mcleod, H. (1986). Computers and children of primary school age: issues and questions. *British Journal of Educational Technology*, *2*, 133–144.

Siann, G., Durndell, A., Mcleod, H., & Glissov, P. (1988). Stereotyping in relation to gender gap participation in computers. *Educational Research*, *30*, 98–100.

Teasley, S. (1995). The role of talk in children's peer collaboration. *Developmental Psychology*, *31*(2), 207–220.

Trowbridge, D. (1987). An investigation of groups working at the computer. In K. Berge, K. Pzdek, & W. Banks (Eds.), *Applications of Cognitive Psychology Problem Solving, Education and Computing*. Hillsdale, NJ: Lawrence Erlbaum Associates.

Underwood, G., & Underwood, J. (1997). Children's interactions and learning outcomes with interactive books. In *Proceedings of Computer Assisted Learning 1997* (University of Exeter, March).

Underwood, G., McCafferey, M., & Underwood, J. (1990). Gender differences in a computer based language task. *Educational Research*, *32*, 44–49.

Underwood, G., Underwood, J., & Turner, M. (1993). Children's thinking during collaborative computer based problem solving. *Educational Psychology*, *13*(3/4), 345–357.

Underwood, G., Jindhal, N., & Underwood, J. (1994). Gender differences and effects of co-operation in a computer based language task. *Educational Research*, *36*(1), 63–74.

Webb, N.M. (1984). Microcomputer learning in small groups: Cognitive requirements and group processes. *Journal of Educational Psychology*, *74*, 475–484.

Wild, M., & Braid, P. (1996). Children's talk in cooperative groups. *Journal of Computer Assisted Learning*, *12*, 216–231.

Yelland, N.J. (1993). Young children learning with LOGO: An analysis of strategies and interactions. *Journal of Educational Computing Research*, *9*(4), 465–486.

Yelland, N.J. (1994). The strategies and interactions of young children in LOGO tasks. *Journal of Computer Assisted Learning*, *10*, 33–49.

Yelland, N.J. (1995). Collaboration and learning with LOGO: Does gender make a difference? In *Proceedings of Computer Supported Collaborative Learning 95* (Bloomington, IN, October 17–20) (pp. 397–401).

## Chapter 6

Aronson, E., Bridgeman, D.L., & Geffner, R. (1978). The effects of a cooperative classroom structure on student behavior and attitudes. In D. Bar-Tal, & L. Saxe (Eds.), *Social Psychology of Education: Theory and Research*. Washington, DC: Wiley.

Bargh, J.A., & Schul, Y. (1980). On the cognitive benefits of teaching. *Journal of Educational Psychology*, *72*, 593–604.

Benware, C.A., & Deci, E.L. (1984). Quality of learning with an active versus passive motivational set. *American Educational Research Journal*, *21*, 755–765.

Bielaczyc, K., Pirolli, P.L., & Brown, A.L. (1995). Training in self-explanation and self-regulation strategies: Investigating the effects of knowledge acquisition activities on problem solving. *Cognition and Instruction*, *13*, 221–252.

Chi, M.T.H., & VanLehn, K. (1991). The content of self-explanations. *Journal of the Learning Sciences*, *1*, 69–105.

Chi, M.T.H., Bassok, M., Lewis, M.W., Reimann, P., & Glaser, R. (1989). Self-explanations: How students study and use examples in learning to solve problems. *Cognitive Science*, *13*, 145–182.

Chi, M.T.H., de Leeuw, N., Chiu, M.-H., & LaVancher, C. (1994). Eliciting self-explanations improves understanding. *Cognitive Science*, *18*, 439–477.

Clark, H.H., & Marshall, C.R. (1981). Definite reference and mutual knowledge. In A.K. Joshi, B. Webber, & I. Sag (Eds.), *Linguistic Structure and Discourse Setting* (pp. 10–63). Cambridge, MA: Cambridge University Press.

Clark, H.H., & Schober, M. (1989). Understanding by addressees and overhearers. *Cognitive Psychology*, *21*, 211–232.

Cloward, R.D. (1967). Studies in tutoring. *Journal of Experimental Education*, *36*, 14–25.

Cohen, P.R. (1978). On knowing what to say: Planning speech acts. Unpublished doctoral dissertation, Department of Computer Science, University of Toronto.

Dayer, P. (1996). Le tutorat virtuel: Apprendre en construisant un didacticiel. Unpublished graduate thesis, Faculty of Education and Psychology, University of Geneva.

Fantuzzo, J.W., Riggio, R.E., Connelly S., & Dimeff, L.A. (1989). Effects of reciprocal peer tutoring on academic achievement and psychological adjustment: A component analysis. *Journal of Educational Psychology*, *81*, 173–177.

Ferguson-Hessler, M.G.M., & de Jong, T. (1990). Studying physics texts: Differences in study processes between good and poor solvers. *Cognition and Instruction*, *7*, 41–54.

Goodlad, S., & Hirst, B. (1989). *Peer Tutoring—A Guide to Learning by Teaching*. London: Kogan.

Halliday, D., & Resnick, R. (1981). *Fundamentals of Physics* (2nd edn.). New York: Wiley.

Kafai, Y., & Harel, I. (1991a). Children learning through consulting: When mathematical ideas, knowledge of programming and design, and playful discourse are intertwined. In I. Harel, & S. Papert (Eds.), *Constructionism* (pp. 111–140). Norwood, NJ: Ablex.

Kafai, Y., & Harel, I. (1991b). Learning through design and teaching: Exploring social and collaborative aspects of contructionism. In I. Harel, & S. Papert (Eds.), *Constructionism* (pp. 85–110). Norwood, NJ: Ablex.

Knight, G.P., & Bohlmeyer, E.M. (1990). Cooperative learning and achievement: Methods for assessing causal mechanisms. In S. Sharan (Ed.), *Cooperative Learning—Theory and Research* (pp. 1–22). New York: Praeger.

Krauss, R.M., & Weinheimer, S. (1964). Changes in reference phrases as a function of the frequency of usage in social interaction: A preliminary study. *Psychonomic Science*, *1*, 113–114.

Mitchell, T.M., Keller, R.M., & Kedar-Cabelli, S.T. (1986). Explanation-based generalization: A unifying view. *Machine Learning*, *1*, 47–80.

Palincsar, A.S., & Brown, A.L. (1984). Reciprocal teaching of comprehension-fostering and comprehension-monitoring activities. *Cognition and Instruction*, *1*, 117–175.

Pirolli, P., & Recker, M. (1994). Learning strategies and transfer in the domain of programming. *Cognition and Instruction*, *12*, 235–275.

Preier, M. (1996). Lernen beim Erklären: Was lernt man selbst, was der Zuhörer? [Learning by explaining: What learns the explainer and what the listener?] Unpublished graduate thesis, Department of Psychology, University of Freiburg.

Renkl, A. (1995). Learning for later teaching: An exploration of mediational links between teaching expectancy and learning results. *Learning and Instruction*, *5*, 21–36.

Renkl, A. (1996). Lernen durch Erklären—oder besser doch durch Zuhören? [Learning by explaining—or better by listening?] *Zeitschrift für Entwicklungspsychologie und Pädagogische Psychologie*, *28*, 148–168.

Renkl, A. (1997a). Learning from worked-out examples: A study of interindividual differences. *Cognitive Science*, *21*, 1–29.

Renkl, A. (1997b). Lernen durch Erklären: Was, wenn Rückfragen gestellt werden? [Learning by explaining: What happens if questions are posed?] *Zeitschrift für Pädagogische Psychologie*, *11*, 41–51.

Schwartz, D.L. (1995). The emergence of abstract representations in dyad problem solving. *Journal of the Learning Sciences*, *4*, 321–354.

Taylor, J., & Carletta, J. (1994). Why we can get away with simple belief models for cooperating agents. In A. Ram, & K. Eiselt (Eds.), *Proceedings of the Sixteenth Annual Conference of the Cognitive Science Society* (pp. 858–863). Hillsdale, NJ: Lawrence Erlbaum Associates.

Towle, A. (1989). *Modern Biology*. New York: Holt, Rinehart & Winston.

VanLehn, K., Jones, R.M., & Chi, M.T.H. (1992). A model of the self-explanation effect. *Journal of the Learning Sciences*, *2*, 1–59.

Webb, N.M. (1989). Peer interaction and learning in small groups. *International Journal of Educational Research*, *13*, 21–39.

## Chapter 7

Baker, M.J. (1989). Negotiated tutoring: An approach to interaction in intelligent tutoring systems. Unpublished Ph.D. thesis, Institute of Educational Technology, The Open University, Milton Keynes.

Baker, M.J. (1994). A model of negotiation in teaching–learning dialogues. *Journal of Artificial Intelligence in Education*, *5*(2), 199–254.

Baker, M.J. (1995). Negotiation in collaborative problem-solving dialogues. In R.-J. Beun, M. J. Baker, & M. Reiner (Eds.), *Dialogue and Instruction*, (pp. 39–55). Berlin: Springer-Verlag.

Bunt, H.C. (1989). Information dialogues as communicative action in relation to partner modelling and information processing. In M.M., Taylor, F. Néel, & D. G. Bouwhuis (Eds.), *The Structure of Multimodal Dialogue* (pp. 47–74). New York: North-Holland/Elsevier Sciences Publishers.

Bunt, H.C. (1995). Dialogue control functions and interaction design. In R.J. Beun, M.J. Baker, & M. Reiner (Eds.), *Dialogue and Instruction, Modeling Interaction in Intelligent Tutoring Systems* (Proceedings of the NATO Advanced Research Workshop on Natural Dialogue and Interactive Student Modeling) (pp. 197–214). Berlin: Springer.

Clark, H.H., & Schaeffer, E.F. (1989). Contributing to discourse. *Cognitive Science. 13*, 259–294.

Cypher, A. (1993). *Watch What I Do: Programming by Demonstration.* Cambridge, MA: MIT Press.

Dillenbourg, P. (1996). Some technical implications of distributed cognition on the design of interactive learning environments. *Journal of Artificial Intelligence in Education, 7* (2), 161–179.

Dillenbourg, P., & Baker, M. (1996). Negotiation spaces in human–computer collaborative learning. In *Proceedings of the International Conference on Cooperative Systems (COOP'96)* (Juan-Les-Pins, France, June 12–14).

Dillenbourg, P., & Self, J.A. (1992). A computational approach to socially distributed cognition. *European Journal of Psychology of Education, 3*(4), 353–372.

Dillenbourg, P., & Traum, D. (1997). The role of a whiteboard in a distributed cognitive system. *Swiss Workshop on Collaborative and Distributed Systems* (Lausanne, May 1).

Goldberg, D.E. (1989). *Genetic Algorithms in Search, Optimisation, and Machine Learning.* Reading, MA: Addison-Wesley.

Grice, H.P. (1975). Logic and conversation. In P. Cole, & J. Morgan (Eds.), *Syntax and Semantics 3: Speech Acts* (pp. 41–58). New York: Academic Press.

Hayes-Roth, F. (1983). Using proofs and refutations to learn from experience. In T.M. Mitchell et al. (Eds.), *Machine Learning: An A.I. Approach* (pp. 221–240). New York: Springer-Verlag.

Hobbs, J.R. (1982). Towards an understanding of coherence in discourse. In W.G. Lehnert, & M.H. Ringle (Eds.), *Strategies for Natural Language Processing* (pp. 223–244). Hillsdale, NJ: Lawrence Erlbaum Associates.

Lakatos, I. (1984). *Preuves et refutations.* Paris: Editions Hermann.

Levinson, S.C. (1983). *Pragmatics.* Cambridge: Cambridge University Press.

Mann, W.C., & Thompson, S.A. (1988). Rhetorical structure theory: Toward a functional theory of text organisation. *Text, 8*(3), 243–281.

Mephu Nguifo, E. (1994). Galois lattice: A framework for concept learning. Design, evaluation and refinement. In *Proceedings of 6th International Conference on Tools with Artificial Intelligence* (New Orleans, LA, November 6–9) (pp. 461–467). Atlanta, GA: IEEE Press.

Mephu Nguifo, E. (1995). Multi-level concept modelling for SBL. In D. Fensel et al. (Eds.), *Proceedings of MLNet (ECML). Workshop on Knowledge Level and Machine Learning,* Heraklion (Crete, Greece, April 28–29). (pp. 81–90). University of Karlsruhe.

Mephu Nguifo, E. (1997). An interactive environment for dynamic control of machine learning systems. In M.J. Smith, G. Salvendy, and R.J. Koubek (Eds), *Design of Computing Systems: Social and Ergonomic Considerations* (Vol. 21B, pp. 31–34). (Proceedings of the 7th International Conference on Human Computer Interaction, San Franscisco, CA, August 24–29). Amsterdam: Elsevier Science.

Michalski, R.S. (1993). Inferential theory of learning as a conceptual basis for multistrategy learning. *Machine Learning. 11*(2/3): 111–151.

Minsky, M. (1987). *The Society of Mind.* London: William Heinemann.

Mitchell, T.M., Utgoff, P.E., & Banerji , R. (1983). Learning by experimentation: Acquiring and refining problem solving heuristics. In R.S. Michalski, J.G. Carbonell, & T.M. Mitchell

(Eds.), *Machine Learning: An Artificial Intelligence Approach* (Vol. I, pp. 163–190). New York: Springer-Verlag.
Moeschler, J. (1985). *Argumentation et conversation: Eléments pour une analyse pragmatique du discours*. Paris: Hatier-Crédif.
Morik, K., Wrobel, S., Kietz, J.U., & Emde, W. (1993). *Knowledge Acquisition and Machine Learning—Theory, Methods and Applications*. London: Academic Press.
Perrault, R. (1990). An application of default logic to speech act theory. In P.R. Cohen, J. Morgan, and M.E. Pollack (Eds.), *Intentions in Communication*, Cambridge, MA: MIT Press.
Roschelle, J., & Teasley, S. (1995). The construction of shared knowledge in collaborative problem solving. In C.E. O'Malley (Ed.), *Computer-Supported Collaborative Learning*. Heidelberg: Springer-Verlag.
Rosenbloom, P.S., Laird, J.E. & Newell, A. (1993). *The Soar Papers: Readings on Integrated Intelligence*. Cambridge, MA: MIT Press.
Salomon, G. (Ed.) (1993). *Distributed Cognitions. Psychological and Educational Considerations* (pp. 111–138). Cambridge, MA: Cambridge University Press.
Sanders, T.J.M., Spooren, W.P.M., & Noordman, L.G.M. (1992). Towards a taxonomy of coherence relations. *Discourse Processes*, *15*, 1–35.
Schwartz, D.L. (1995). The emergence of abstract dyad representations in dyad problem solving. *Journal of the Learning Sciences*, *4*(3), 321–354.
Wrobel, S. (1994). *Concept Formation and Knowledge Revision*, Dordrecht, The Netherlands: Kluwer Academic.
Yoshida, K., & Motoda, H. (1996). Automated user modelling for intelligent interface. *International Journal of Human–Computer Interaction*, *8*(3), 237–258.

## Chapter 8

Aebli, H. (1980). *Denken: Das Ordnen des Tuns. Band 1: Kognitive Aspekte der Handlungstheorie*. Stuttgart: Klett-Kotta.
Anderson, J.R., Boyle, A.T., Corbett, A., & Lewis, M. (1990). Cognitive modeling and intelligent tutoring. *Artificial Intelligence*, *42*, 7–51.
Aronson, E., Balney, N., Stephan, C., Sikes, J., & Snapp, M. (1978). *The Jigsaw Classroom*. Beverley Hills, CA: Sage.
Barenholz, H., & Tamir, P. (1992). A comprehensive use of concept in design, instruction and assessment. *Research in Science and Technological Education*, *10*(1), 37–52.
Beller, S., & Hoppe, H.U. (1993). Deductive error reconstruction and classification in a logic programming framework. In P. Brna, S. Ohlsson, & H. Pain (Eds.), *Proceedings of the Sixth World Conference on Artificial Intelligence in Education* (pp. 433–440). Charlottesville, VA: Association for the Advancement of Computing in Education.
Clancey, W. (1992). Representations of knowing: In defense of cognitive apprenticeship. *Journal of Artificial Intelligence in Education*, *3*, 139–168.
Dewan, P., & Choudhary, R. (1991). Flexible user interface coupling in a collaborative system. In *Proceedings of the International Conference on Computer–Human Interaction* (pp. 41–48). New Orleans, LA: ACM Press.
Doignon, J.-P., & Falmagne, J.-C. (1985). Spaces for the assessment of knowledge. *International Journal of Man–Machine Studies*, *23*, 175–196.
Doise, W., & Mugny, G. (1984). *The Social Development of the Intellect*. Oxford: Pergamon.

Greenberg, S., Roseman, M., Webster, D., & Bohnet, R. (1992). Human and technical factors of distributed group drawing tools. *Interacting with Computers, 4*, 364–392.

Hoppe, H.U. (1994). Deductive error diagnosis and inductive error generalization for intelligent tutoring systems. *Journal of Artificial Intelligence in Education, 5*, 27–49.

Hoppe, H.U. (1995). Using multiple student modeling to parameterize group learning. In J. Greer (Ed.), *Proceedings of the Seventh World Conference on Artificial Intelligence in Education* (pp. 234–241). Charlottesville, VA: Association for the Advancement of Computing in Education.

Hoppe, H.U., Baloian, N., & Zhao, J. (1993). Computer support for teacher-centered classroom interaction. In *Proceedings of the International Conference on Computers in Education* (Taipei, R.O.C.) (pp. 211–217).

Howe, C., Tolmie, A., & Rodgers, C. (1990). Physics in the primary school: Peer interaction and the understanding of floating and sinking. *European Journal of Psychology of Education, 4*, 459–475.

Howe, C., Tolmie, A., & Rodgers, C. (1992). The acquisition of conceptual knowledge in science by primary school children: Group interaction and the understanding of motion down an incline. *British Journal of Developmental Psychology, 10*, 113–130.

Knight, G.P., & Bohlmeyer, E.M. (1990). Cooperative learning and achievement: Methods for assessing causal mechanisms. In S. Sharan (Ed.), *Cooperative Learning—Theory and Research* (pp. 1–22). New York: Praeger.

McCalla, G.I. (1992). The centrality of student modeling to intelligent tutoring. In E. Costa (Ed.), *New Directions for Intelligent Tutoring Systems* (pp. 107–131). Berlin: Springer.

McCalla, G.I., Greer, J.E., Kumar, V.S., Meagher, P., Collins, J.A., Tkatch, R., & Parkinson, B. (1997). A peer help system for workplace training. In B. du Boulay, & R. Mizoguchi (Eds.), *Proceedings of the Eighth World Conference on Artificial Intelligence in Education* (pp. 183–190). Amsterdam: IOS Press.

Mühlenbrock, M., Tewissen, F., & Hoppe, H.U. (1997). A framework system for intelligent support in open distributed learning environments. In B. du Boulay, & R. Mizoguchi (Eds.), *Proceedings of the Eighth World Conference on Artificial Intelligence in Education* (pp. 191–198). Amsterdam: IOS Press.

Novak, J.D. (1990). Concept mapping: A useful tool for science education. *Journal of Research in Science Teaching, 27*, 937–949.

Novak, J.D., & Gowin, D.B. (1984). *Learning How to Learn*. Cambridge, MA: Cambridge University Press.

Ohlsson, S. (1988). Computer simulation and its impact on educational research and practice. *International Journal of Educational Research, 12*, 5–12.

Paiva, A. (1997). Learner modelling for collaborative learning environments. In B. du Boulay, & R. Mizoguchi (Eds.), *Proceedings of the Eighth World Conference on Artificial Intelligence in Education* (pp. 215–222). Amsterdam: IOS Press.

Palincsar, A.S., & Brown, A.L. (1984). Reciprocal teaching of comprehension-fostering and comprehension-monitoring activities. *Cognition and Instruction, 1*, 117–175.

Ploetzner, R. (1995). The construction and coordination of complementary problem representations in physics. *Journal of Artificial Intelligence in Education, 6*, 203–238.

Ploetzner, R., Fehse, E., Spada, H., Vodermaier, A., & Wolber, D. (1996a). Physiklernen mit modellgestützt konstruierten Begriffsnetzen und durch Problemlösen zu zweit [Learning physics by means of model-based concept maps and collaborative problem solving]. *Zeitschrift für Entwicklungspsychologie und Pädagogische Psychologie, 28*, 270–293.

Ploetzner, R., Hoppe, H.U., Fehse, E., Nolte, C., & Tewissen, F. (1996b). Model-based design of activity spaces for collaborative problem solving and learning. In P. Brna, A. Paiva, & J.

Self (Eds.), *Proceedings of the European Conference on Artificial Intelligence in Education* (pp. 372–378). Lisbon: Colibri.
Ploetzner, R., Fehse, E., Hermann, F., & Kneser, C. (1997). Modeling the knowledge-based exchange of information during collaborative problem solving on the basis of deductive self-diagnosis. In B. du Boulay, & R. Mizoguchi (Eds.), *Proceedings of the Eighth World Conference on Artificial Intelligence in Education* (pp. 223–230). Amsterdam: IOS Press.
Roschelle, J., & Teasley, S.D. (1995). The construction of shared knowledge in collaborative problem solving. In C. O'Malley (Ed.), *Computer Supported Collaborative Learning* (pp. 67–97). Berlin: Springer.
Self, J. (1994). The role of student models in learning environments. *IEICE Transactions on Information and Systems, E77-D*(1), 3–8.
Simon, H.A. (1969). *The Sciences of the Artificial*. Cambridge, MA: MIT Press
Smith, R.B., O'Shea, T., O'Malley, C., Scanlon, E., & Taylor, J. (1991). Preliminary experiments with a distributed, multimedia, problem solving environment. In J.M. Bowers, & S. D. Benford (Eds.), *Studies in Computer Supported Cooperative Work* (pp. 31–48). Amsterdam: Elsevier Science.
Sterling, L., & Shapiro, E. (1994). *The Art of PROLOG* (2nd edn.). Cambridge, MA: MIT Press.
Suchman, L.A. (1987). *Plans and Situated Actions. The Problem of Human–Machine Communication*. Cambridge, MA: Cambridge University Press.
Suthers, D., Weiner, A., Connelly, J., & Paolucci, M. (1995). Belvedere: Engaging students in critical discussion of science and public policy issues. In J. Greer (Ed.), *Proceedings of the Seventh World Conference on Artificial Intelligence in Education* (pp. 266–273). Charlottesville, VA: Association for the Advancement of Computing in Education.
VanLehn, K., Jones, R.M., & Chi, M.T.H. (1992). A model of the self-explanation effect. *Journal of the Learning Sciences, 2*, 1–59.
Webb, N.M. (1989). Peer interaction and learning in small groups. *International Journal of Education Research, 13*, 21–39.
Zhao, J., & Hoppe, H.U. (1994). Supporting flexible communication in heterogeneous multi-user environments. In *Proceedings of the Fourteenth IEEE International Conference on Distributed Computing Systems* (pp. 442–449). Los Alamitos, CA: IEEE Computer Society Press.

## Chapter 9

Allwood, J., Nivre, J., & Ahlsén, E. (1991). *On the Semantics and Pragmatics of Linguistic Feedback* (Gothenburg Papers in Theoretical Linguistics No. 64). Sweden: University of Gothenburg, Department of Linguistics.
Aytes, K. (1995). Comparing collaborative drawing tools and whiteboards: An analysis of the group process. *Computer Supported Cooperative Work, 4*, 51–71.
Bennett, R., & Kirstein, P.T. (1997). Technical innovations deployed by the MERCI Project. In *Proceedings of the Networkshop 25* (Belfast, March) (pp. 181–189).
Bygholm, A., & Dirckinck-Holmfeld, L. (1997). Pædagogik i Det virtuelle Læremiljø—Metodiske overvejelser [Pedagogy in the virtual learning environment—methological considerations]. In O. Danielsen (Ed.), *Læring og multimedier* [Learning and Multimedia] (pp. 63–86). Aalborg: Aalborg University Press.

Clark, H.H., & Brennan, S.E. (1991). Grounding in communication. In L.B. Resnick, J. Levine, & S.D. Teasley (Eds.), *Perspectives on Socially Shared Cogniton* (pp. 127–149). Washington, DC: American Psychological Association.

Cockburn, A.J.G., & Greenberg, S. (1993). Making contact: Getting the group communicating with groupware. In *Proceedings of the ACM Conference on Organisational Computing Systems, COOCS '93* (Milpitas, CA) (pp.31–41).

Cole, M., & Engeström Y. (1993). A cultural-historical approach to distributed cognition. In G. Salomon (Ed.), *Distributed Cognitions—Psychological and Educational Considerations* (pp. 1–47). Cambridge: Cambridge University Press.

Daft, R.L., & Lengel, R.H. (1984). Information richness: A new approach to managerial behavior and organisation design. *Research in Organizational Behavior, 6,* 191–233.

Dirckinck-Holmfeld, L. (1990). *Kommunikation på trods og på tværs—Projektpædagogik og datamatkonferencer i fjernundervisning* [Project Pedagogy and Computer Conferencing in Distance Learning]. Picnic-News No. 9. Aalborg: Aalborg University.

Dix, A. (1997). Challenges for cooperative work on the Web: An analytical approach. *Computer Supported Cooperative Work, 6,* 135–156.

Dorcey, T. (1995). CU-SeeMe desktop videoConferencing software. *Connexions, 9*(3). Retrieved November 10, 1998 from the World Wide Web: http://ftp.sunet.se/ftp/pub/conferencing/CU-SeeMe/html/DorceyConnexions.html

Engeström (Ed.), Y. (1987). *Learning by Expanding—An Activity-Theoretical Approach to Developmental Research.* Helsinki: Orienta-Konsultit Oy.

Engeström, Y. (1990). When is a tool? Multiple meanings of artifacts in human activity. In Y. Engeström (Ed.), *Learning, Working and Imagining. Twelve Studies in Activity Theory* (pp. 171–195). Jyväskylässä: Orienta-Konsultit Oy.

Engeström, Y. (1993). Developmental studies of work as a testbench of theory: The case of primary care medical practice. In S. Chaiklin, & J. Lave (Eds.), *Understanding Practice: Perspectives on Activity and Context* (pp. 64–103). Canada: Cambridge University Press.

Engeström, Y. (1995). Innovative organizational learning in medical and legal settings. In L. Martin, K. Nelsen, & E. Tobach (Eds.), *Sociocultural Psychology: Theory and Practice of Doing and Knowing* (pp. 326–356). Cambridge, MA: Cambridge University Press.

Engeström, Y., & Engeström, R. (1986). Developmental work research: The approach and an application in cleaning work. *Nordisk Pedagogik, 1,* 2–15.

Fjuk, A., & Dirckinck-Holmfeld, L. (1996). Problem oriented collaborative distance learning: why so difficult? In B. Dahlbom, F. Ljungberg, U. Nuldén, K. Simon, & C. Sørensen (Eds.), *Proceedings of the 19th Information Systems Research Seminar in Scandinavia, IRIS 19,* Vol. 1.

Fjuk, A., & Dirckinck-Holmfeld, L. (1997). Sammenføyningsarbeid i distribuerte kollektive læreprocesser [Articulation work in collaborative learning]. In O. Danielsen (Ed.), *Læring og multimedier* [Learning and Multimedia] (pp. 145–176). Aalborg: Aalborg University Press.

Galegher, J., & Kraut, R.E. (1990). Technology for intellectual teamwork: Perspectives on research and design. In J. Galegher, R.E. Kraut, & C. Egido (Eds.), *Intellectual Teamwork. Social and Technological Foundations for Cooperative Work* (pp. 1–20). Hillsdale, NJ: Lawrence Erlbaum Associates.

Georgsen, M. (1995). Den anden dimension. Computer-medieret kommunikation og collaborative learning [The second dimension. Computer-mediated communication and collaborative learning]. Thesis, Aalborg University.

Hansen, T. (1995). Erkendelses-deling [Sharing cognition]. Magisterkonferens (thesis), Institute of Psychology, University of Aarhus.

Hansen, T. (in preparation). Cognitive collaboration. Ph.D. project, University of Aarhus and Aalborg University.

Hedegaard, M., Hansen, E., Engeström, Y., & Jensen, U.J. (1985). Fagundervisning—Elevundervisning og udvikling af teoretisk tænkning [Didactics—Teaching and development of theoretical thought]. Psykologisk Skriftserie Aarhus, Institute of Psychology, University of Aarhus.

Heeren, E., & Lewis, R. (1997). Selecting communication media for distributed communities. *Journal of Computer Assisted Learning*, *13*(2), 85–98.

Henri, F. (1995). Distance learning and computer mediated communication: Interactive, quasi-interactive or monologue? In C. O'Malley (Ed.), *Computer Supported Collaborative Learning* (pp. 145–161). Berlin: Springer-Verlag.

Hutchins, E. (1990). The technology of team navigation. In J. Galegher, R.E. Kraut, & C. Egido (Eds.), *Intellectual Teamwork. Social and Technological Foundations of Cooperative Work* (pp. 191–220). Hillsdale, NJ: Lawrence Erlbaum Associates.

Hutchins, E. (1993). Learning to navigate. In S. Chaiklin, & J. Lave (Eds.), *Understanding Practice: Perspectives on Activity and Context* (pp. 35–63). Canada: Cambridge University Press.

Ihde, D.A. (1975). A phenomenology of man–machine relations. In R. Feinberg, & E. Rosemont (Eds.), *Work, Technology and Education* (pp. 186–203). Urbana, IL: University of Illinois Press.

Illeris, K. (1981). *Modkvalificeringens Pædagogik* [Emancipatoric Pedagogy]. Denmark: Unge Pædagoger.

Leontiev, A.N. (1978). *Activity, Consciousness, Personality*. Englewood Cliffs, NJ: Prentice Hall.

Lewis, R. (1995). Editorial: Professional learning. *Journal of Computer Assisted Learning*, *11*(4), 193–195.

Lewis, R. (1997). An activity theory framework to explore distributed communities. *Journal of Computer Assisted Learning*, *13*(4), 210–218.

Linard, M. (1995). New debates on learning support. *Journal of Computer Assisted Learning*, *11*(4), 239–252.

Lubich, H. (1995). *Towards a CSCW Framework for Scientific Cooperation in Europe*. Berlin: Springer-Verlag.

Mammen, J. (1989). The elements of psychology. In N. Engelsted, M. Hedegaard, B. Karpatschof, & A. Mortensen (Eds.), *The Societal Subject* (pp. 29–44). Aarhus: Aarhus University Press.

Nardi, B.A. (Ed.). (1996). *Context and Consciousness—Activity Theory and Human–Computer Interaction*. Cambridge, MA: MIT Press.

Reinhard, W., Schweitzer, J., Volksen, G., & Weber, M. (1994). CSCW tools: Concepts and architectures. *IEEE Computer*, *27*(5), 28–36.

Rugelj, J. (1992). Simple distributed computer supported multimedia environment for education. In *Proceedings of the East–West Conference on Computers in Education* (Moscow) (pp. 343–345).

Rugelj, J. (1996). Groupware tools for collaboration and information sharing in education. In *Proceedings of the International Conference on Multimedia Technology and Digital Telecommunication Services, ICOMT '96* (Budapest, Hungary) (pp. 139–144).

Rugelj, J. (1997). Teleinformatics in education: Globalization and local obstacles. In P. Enslow, P. Desrochers, & I. Bonifacio (Eds.), *Global Networking '97* (pp. 317–320). Amsterdam: IOS Press.

Säljö, R. (1996). Mental and physical artefacts in cognitive practices. In P. Reimann, & H. Spada (Eds.), *Learning in Humans and Machines* (pp. 83–96). Oxford: Pergamon.
Star, S.L. (1989). The structure of ill-structured solutions: Boundary objects and heterogeneous distributed problem solving. In M. Huhns, & L. Gasser (Eds.), *Distributed Artificial Intelligence, Vol. 2* (pp. 37–54). London: Pitman.
VMDL (1996). *Final Report of the Virtual Mobility and Distributed Laboratories Project.* Retrieved November 10, 1998 from the World Wide Web: http://www.lancs.ac.uk/users/ktru/vmdl.htm.
von Cranach, M. (1982). The psychological study of goal-directed action: Basic issues. In M. von Cranach, & R. Harre (Eds.), *The Analysis of Action: Recent Theoretical and Empirical Advances* (pp. 35–73). Cambridge: Cambridge University Press.
Vygotsky, L.S. (1978). *Mind in Society. The Development of Higher Psychological Processes.* Cambridge, MA: Harvard University Press.
Watabe, K., Sakata, S., Maeno, K., Fukuoka, H., & Ohmori, T. (1990). Distributed multiparty desktop conferencing system: Mermaid. In *Proceedings of the Conference on Computer Supported Cooperative Work, CSCW '90* (Los Angeles, CA) (pp. 27–38).
Wood, D., Bruner, J.S., & Ross, G. (1976). The role of tutoring in problem solving. *Journal of Child Psychology and Child Psychiatry, 17,* 89–100.

## Chapter 10

Barron, B.J. (1992). Collaborative problem solving: Is team performance greater than what is expected from the most competent member? Unpublished dissertation, Vanderbilt University, Nashville, TN.
Bartlett, F.C. (1932). *Remembering: A Study in Experimental Psychology.* Cambridge: Cambridge University Press.
Bereiter, C., & Scardamalia, M. (1989). Intentional learning as a goal of instruction. In L.B. Resnick (Ed.), *Knowing, Learning, and Instruction: Essays in Honor of Robert Glaser* (pp. 361–392). Hillsdale, NJ: Lawrence Erlbaum Associates.
Bransford, J.D., & Johnson, M.K. (1972). Contextual prerequisites for understanding: Some investigations of comprehension and recall. *Journal of Verbal Learning and Verbal Behavior, 11,* 717–726.
Bransford, J. D., & Nitsch, K.E. (1978). Coming to understand things we could not previously understand. In J.F. Kavanagh, & W. Strange (Eds.), *Speech and Language in the Laboratory, School and Clinic.* Cambridge, MA: MIT Press.
Csikszentmihalyi, M. (1990). *Flow: The Psychology of Optimal Experience.* New York: Harper & Row.
Dennett, D.C. (1987). *The Intentional Stance.* Cambridge, MA: MIT Press.
Dunnette, M.D., Campbell, J., & Jaastad, K. (1963). The effect of group participation on brainstorming effectiveness for two industrial samples. *Journal of Applied Psychology, 47,* 30–37.
Eco, U. (1984). *Semiotics and the Philosophy of Language.* Bloomington, IN: Indiana University Press.
Fox, D.J., & Lorge, I. (1962). The relative quality of decisions written by individuals and by groups as the available time for problem solving is increased. *Journal of Social Psychology, 57,* 227–242.
Gibson, E. (1969). *Principles of Perceptual Learning and Development.* New York: Meredith Corporation.

Hagel, J., & Armstrong, A.G. (1997). *Net Gain: Expanding Markets Through Virtual Communities*. Boston, MA: Harvard Business School Press.

Hastie, R. (1983). Experimental evidence on group accuracy. In B. Grofman, & G. Owen (Eds.), *Information Pooling and Group Decision Making* (pp. 129–157). Greenwich, CT: JAI Press.

Hill, G.W. (1982). Group versus individual performance: Are $N + 1$ heads better than one? *Psychological Bulletin*, *91*, 517–539.

Hoover-Dempsey, K.V., & Jones, K.P. (1996). Parental perceptions of appropriate roles and role activities in children's education. *Paper presented at the International Roundtable on Families, Schools, Communities and Children's Learning* (New York, April).

Kelly, H.H., & Thibaut, J.W. (1969). Group problem solving. In G. Lindzey, & E. Aronson (Eds.), *The Handbook of Social Psychology* (2nd edn., pp. 1–101). Reading, MA: Addison-Wesley.

Krauss, R.M., & Glucksberg, S. (1969). The development of communication: Competence as a function of age. *Child Development*, *40*, 255–266.

Laughlin, P.R., & Futoran, G.C. (1985). Collective induction: Social combination and sequential transition. *Journal of Personality and Social Psychology*, *48*, 608–613.

Marx, K. (1939/1973). *Grundrisse*. (M. Nicolaus, Trans.) New York: Random House.

McClellan, C. (1971). Feuding and warfare among Northwestern Athapaskans. In A. McFadyen Clark (Ed.), *Proceedings: Northern Athapaskan Conference. National Museum of Man Mercury Series Canadian Ethnology Service* (Vol. 1, Paper No. 27).

Norman, D.A. (1993). *Things That Make Us Smart: Defending Human Attributes in the Age of the Machine*. New York: Addison-Wesley.

Robertson, S.P., Zachary, W., & Black, J.B. (Eds.) (1990). *Cognition, Computing, and Cooperation*. Norwood, NJ: Ablex.

Roschelle, J., & Teasley, S.D. (1995). The construction of shared knowledge in collaborative problem solving. In C.E. O'Malley (Ed.), *Computer-Supported Collaborative Learning* (pp. 69–97). Berlin: Springer-Verlag.

Schwartz, D.L. (1995). The emergence of abstract representations in dyad problem solving. *Journal of the Learning Sciences*, *4*, 321–354.

Schwartz, D.L., & Black, T. (1999). Inferences through imagined actions: Knowing by simulated doing. *Journal of Experimental Psychology: Learning, Memory and Cognition*, *25*, 1–21.

Schwartz, D.L., & Bransford, J.D. (1996). Experiences that energize. Unpublished manuscript, Vanderbilt University, Nashville, TN.

Schwartz, D.L., & Bransford, J.D. (in press). A time for telling. *Cognition and Instruction*.

Slavin, R.E. (1983). *Cooperative Learning*. New York: Longman.

Turro, I., Porter, R.H., & Picard, M. (1994). Olfactory cues mediate food selection by young chicks. *Physiology and Behavior*, *55*, 761–767.

Vye, N.J., Schwartz, D.L., Bransford, J.D., Barron, B.J., Zech, L., & Cognition and Technology Group at Vanderbilt (1998). SMART environments that support monitoring, reflection, and revision. In D. Hacker, J. Dunlosky, & A. Graesser (Eds.), *Metacognition in Educational Theory and Practice*. (pp. 305–346) Mahwah, NJ: Lawrence Erlbaum Associates.

Weizenbaum, J. (1966). ELIZA—A computer program for the study of natural language communication between man and machine. *Communications of the ACM*, *9*, 36–44.

# Index

activity theory *see* cultural–historical activity theory (CHAT)
agents 5, 10–11, 65–7, 79, 145
  dialogue/machine learning distinction 139, 140–1, 142
  grounding 33–4, 47, 48–9, 50–2, 53
  knowledge transformations 122–3, 133
  *see also* multi-agent learning; productive agency
AI *see* artificial intelligence
analytic models 147–68
apprenticeship 3, 27–8
appropriation
  grounding 16, 31–2, 41, 44–6, 57, 61–3
  production dialectic 198–9, 207–8, 209–10, 217
argumentation 23–4, 123, 126, 135, 164
artificial intelligence (AI) 4, 64–80, 142, 199, 200, 204–7, 217
  *see also* distributed artificial intelligence (DAI)
asymmetry 8, 10, 15, 24, 153

C-CHENE 33, 36–7, 52–62
CASCADE 105, 109–10
CHAT *see* cultural–historical activity theory
children 21–8, 31–2, 41–4, 88, 89, 185
CKC *see* collaborative knowledge construction
cognitive load 6, 14–15, 61, 77, 120, 122
collaborative knowledge construction (CKC) 169–72, 177–8, 183, 186, 189–95
communication
  breakdowns 217
  cognitive simulation model 150–3
  design issues 162–3
  foraging test 92–3, 95–100, 101, 102
  grounding 39, 44
  group performance 88
  multi-robot systems 83, 86–7, 91–2
  rules 218
  synchronicity 12
  telematics 162–3, 169–96, 206
  *see also* computer-mediated communication; dialogue; language
computational model 4–5, 19, 38, 77, 122, 200
computer modeling 199–201, 202, 210, 216–18
computer-mediated communication (CMC) 28, 31, 35, 36–8, 45, 63, 171
  *see also* interfaces
computer-supported collaborative learning (CSCL) 20–30, 81, 87–90
  analytic models 5, 148–9
  grounding 33, 36–7, 52–62
  group memory 3
  group size 2, 19
  interaction monitoring 17
  monitoring 8–9
  multi-robot systems comparison 90–2
  rules 8
  telematics 169, 171, 176–9
computer-supported cooperative/collaborative work (CSCW) 12, 50, 145, 171, 179–83, 188–9
concept maps 164–5
conferencing, telematics 171, 174–6, 177–9, 181, 182, 195
conflict 15, 21–4, 76, 87
  *see also* disagreement
context 27, 127, 139, 140
cooperation
  collaboration distinction 11, 21
  complementarity 160
  peer interaction 88
  productive agency 198, 200, 202
  robots 72, 74, 82, 86
coordination 21, 163, 169, 172, 204
  cognitive simulation model 152

coordination (*cont'd*)
grounding 57, 61
multi-agent learning 70, 71, 73
CSCL *see* computer-supported collaborative learning
CSCW *see* computer-supported cooperative/collaborative work
cultural–historical activity theory (CHAT)
grounding 31, 32–3, 40–1, 43, 45, 51, 62
telematics 169, 171, 183–6, 193–4
culture 3, 32, 41, 183, 202–3

DAI *see* distributed artificial intelligence
derivations 124, 126, 144
diagnosis, cognitive simulation model 34, 73, 150–1, 152, 158, 161
dialogue
artificial agents 79
C-CHENE example 36–7, 58–60
explanation 120
joint problem space 26
machine learning comparison 4, 122–46, 210
negotiability 13–14
shared understanding 76
disagreement 13, 25, 39
distributed artificial intelligence (DAI) 5, 10, 18–19, 199, 210
misunderstanding 76, 78
multi-agent learning 64, 65–7, 68, 75
division of labour 11, 14, 15, 70–1, 75, 77
division mechanisms 70–2, 73, 74, 75, 77–8, 80

explanation 11, 15, 77–8, 103–20, 149–50, 211
*see also* self-explanation

feedback 29, 34–5, 38, 57, 116–18, 157, 211

gender 28, 29, 89–90
generalisation 2, 4, 77
machine learning operators 125, 128, 131, 133, 135–8, 140
goals 10–11, 21, 189–92, 217
grounding 47, 48, 49, 53, 61, 62
multi-agent learning 68, 71, 72, 75
grounding 3–4, 9, 13, 14, 27, 31–63
appropriation 16
CKC 186
reification 163
transformations 143
groups
activity model 186–8
formation 7, 158, 178, 192
gender 89–90
model-based support 156–62, 168
multi-agent learning 68, 72
size 2–3, 7, 87, 182, 200

heuristics 134, 138, 217
human–human interaction
model-based group support 158, 168
multi–agent learning 64, 65, 75–9, 80
robot–robot comparison 81–102
human–tool interaction 173–5, 183–4

identity 202–3
ILE *see* interactive learning environments
induction 14, 18, 135, 136–8
intelligent computer-supported collaborative learning (I-CSCL) 148
intelligent tutoring systems (ITS) 148, 162
intention 201–2, 205–6, 207–8
interaction 2, 3, 6–7, 8–9, 11–14, 15
appropriation/production distinction 209
C-CHENE example 36–7, 56–61
computer-based collaborative learning 20, 27–8, 87, 88
conferencing tools 178–9, 195
effects 16–17, 18
explanation 103, 104, 118
grounding 32, 33–4, 35, 39, 45–6, 62
group 156–7, 158, 168
human–tool 173–5, 183–4
knowledge transformation 134–5
levels of communicative function 189–90
machine learning systems 144
modeling 147
multi-agent learning 68, 69, 70–1, 72–6, 78, 80
multi-robot systems 82–3, 91–2
negotiation 123, 126
non-verbal 26, 44, 79
shared meaning 203
shared workspace 156, 164–5, 168, 171–2
socially constructed processes 24–5
*see also* communication; peer facilitation
interactive learning environments (ILE) 157–8, 162

interfaces 8, 36–7, 52–61, 144, 161–2, 170
internalisation 15, 18, 24, 79, 141
intersubjectivity 33, 42–3, 46
ITS *see* intelligent tutoring systems

joint problem space (JPS) 25–6

knowledge 8, 9, 10, 16, 25, 166
  appropriation 198–9, 207–8, 217
  cognitive models 147, 149, 150–2, 156, 159–60, 161
  dialogue/machine learning comparison 4, 122–46, 210
  organizational 69–70
  self-explanation 103, 107, 109–10, 112
  structured 214–16
  teaching 118–19
  transformations 123, 127–34, 136–7, 139–41, 143
  zone of proximal development 185–6
  *see also* collaborative knowledge construction (CKC)

language 26, 41, 78, 180, 210–13
  disambiguation 163-4
  grounding 32–3, 44–7, 46
  structured understanding 214–15
  *see also* linguistics; meaning; semiotic tools
LEGAL 144–5
linguistics 13, 79, 199, 210–11
logico-semantic relations 127

machine learning (ML)
  dialogue operator comparison 4, 18–19, 122–46, 210
  multi-agent learning 64–5, 67–75, 80
material tools 31, 41, 45, 50–1, 54–5, 57–62, 183
meaning 42, 79, 139–40, 202–4, 210–12
media 28, 35, 172–3, 174–5
mediation 27, 40–1, 45, 50
meta-cognition 134, 151, 152–3
misunderstanding 13–14, 64, 76, 78–9, 80, 203
modeling 4, 12, 16, 18, 147–68
  C-CHENE example 52–62
  computers 199–201, 202, 210, 216–18
  explanation 120
  human activity 184–5, 186–8
  joint intentions 74
  multi-robot systems 82–3
  regulation 77
  self-explanation 109–10
motivation 87, 91, 180
  multi-robot systems 81, 82–3, 91
  productive agency 198, 199, 207–10, 216, 217
multi-agent learning 5, 13, 16, 18, 64–80
  dialogue/machine learning operators comparison 139, 142–3
  human–human interaction 64, 65, 75–80
  robot–robot interaction 64–5, 67–75, 80
multiple student modeling 8, 147, 149, 158–62, 168
multiplication mechanisms 68–70, 72, 73, 75, 76, 80
mutual refinement 126–7, 130, 132

negotiation 9, 13, 24, 25–6, 35, 143
  dialogue operators 135, 138
  knowledge 123, 126
  multi-agent learning 74, 79
  shared workspaces 164
  telematics 189, 190, 195
non-verbal interaction 26, 44, 79

on-line support 156–62, 168

peer facilitation 21–4, 27, 28–30, 88
Piaget, Jean
  conflict 15, 21–2, 87, 90
  dialogue 4
  logical reasoning 24
  meaning 203
problem-solving 6, 21, 25–6, 28
  cognitive models 147, 148, 149–56
  dialogue relations 123, 126, 127–8, 130–4
  explanation 105, 113, 117
  multi-robot systems method 92–101
  mutual refinement 126, 130
  pedagogy paradigm 176–7
  self-explanation 105–7, 109–10
production, appropriation dialectic 209–10, 217
productive agency 197–218
PROLOG 152, 153, 161

reasoning 12, 24, 142–3
reciprocal teaching 7, 113–15

reformulation 124, 133
regulation 8, 14, 76–7
  *see also* self-regulation
reification 163
repair 26, 34, 35, 79
richness 167, 173, 190, 193, 195
robots 10, 201, 202
  human–human interaction comparison 81–102
  multi-agent learning 64–5, 67–75, 80
roles 7–8, 74, 149, 150, 181–2, 202–4
rules 5, 8, 77–8, 198, 203–4, 217, 218

scale 2–3, 4, 5, 11
self-explanation 4, 15, 77–8, 103–12, 119–20, 145, 211
self-regulation 8–9, 76–7, 111
semiotic tools 31, 41, 44–6, 50–1, 55–6, 61–2, 183
situation 9–11, 17–18, 44, 49
support
  domain knowledge 166–7
  model-based 149, 156–62
symbolic information processing hypothesis 148
symmetry 9–10, 11, 145
  *see also* asymmetry
synchronicity 12, 14, 18, 164–5, 172

teaching 26–7, 103, 112–19, 209–10
  model-based group support 158, 161–2
  reciprocal 7, 113–15
  telematics 176–9
telematics 148–9, 162–3, 169–96, 206
  *see also* computer-mediated communication
TFs *see* transformation functions
time 84, 101, 193–4
tools 8, 27, 147, 164
  grounding 31–2, 40–7, 48, 50–1, 54–7, 61–3
  telematic 169–96
trace diagrams 92, 96–7, 100–1, 102
transformation functions (TFs) 123, 127–34, 136–7, 139–41, 143
transmutations 123, 124–6, 131, 132–4, 139, 144
transparency 173, 175, 195
Turing test 204–7

understanding 42–3, 113–14, 115–18, 201
  grounding 31, 32, 34–5, 36–7, 44–7, 50–2, 62–3
  interaction mechanisms 75–6
  joint construction of 24–6
  self-explanation 4, 107, 108
  shared knowledge 189, 190, 211
  structured 212–13, 214–15
  *see also* meaning; misunderstanding

VMDL project 180–2
Vygotsky, Lev Semionovich
  CHAT 40, 169, 183
  child development 32
  communitarianism 209
  developmental process 24
  dialogue 4
  peer interaction 27
  ZPD 41–2, 90, 185

whiteboard 145–6, 190
workspace, shared 147, 156, 162–7, 168, 171–2, 190
World Wide Web 209

zone of proximal development (ZPD) 41–2, 43, 46, 47, 90, 185–6